Simon King is a met[eorologist who presents the] weather on BBC TV [and radio. He has also] produced and co-pre[sented podcasts like 10 Minute] *Weather* with Clare Nasir. Simon was previously an operational meteorologist for the Met Office and part of a specialist forecasting unit with the Royal Air Force. Having trained as an RAF reserve officer, he was deployed with the military on operations to the Middle East where he provided crucial weather data and forecasts. He has a postgraduate and undergraduate degree in Meteorology from the University of Reading and is a self-confessed weather nut!

Clare Nasir is a Met Office trained Meteorologist with a BSc in Maths and MSc in Oceanography. She is currently the Channel 5 News weather anchor and also produces and presents the Met Office podcasts. Clare has over 20 years of experience in weather forecasting; operationally and in the media and regularly contributes to weather and climate TV documentaries. She co-presented the acclaimed documentary series *Fierce Earth* for CBBC and has four children books published.

SIMON KING & CLARE NASIR

WHAT DOES RAIN SMELL LIKE?

100 fascinating questions on the wild ways of the weather

First published in the UK in 2019 by 535
This edition published in 2025 by Blink Publishing
An imprint of Bonnier Books UK
5th Floor, HYLO, 105 Bunhill Row,
London, EC1Y 8LZ

Copyright © Simon King and Clare Nasir, 2019

Images: Alamy © p.199, p.201, p.202, p.203, p.207, p.209, p.212, p.219, Getty © p.206, p.208, Met Office © p.316

Illustrations © Clarkevanmeurs Design

All rights reserved.

No part of this publication may be reproduced, stored or transmitted in any form or by any means, electronic, mechanical, photocopying or otherwise, without the prior written permission of the publisher.

The right of Simon King and Clare Nasir to be identified as Authors of this work has been asserted by them in accordance with the Copyright, Designs and Patents Act, 1988.

A CIP catalogue record for this book is available from the British Library.

Hardback ISBN: 9781788702096
Trade Paperback ISBN: 9781788702898
Paperback ISBN: 9781788704663

Also available as an ebook and an audiobook

1 3 5 7 9 10 8 6 4 2

Design and Typeset by seagulls.net
Printed and bound in Great Britain by Clays Ltd, Elcograf S.p.A.

Every reasonable effort has been made to trace copyright holders of material reproduced in this book, but if any have been inadvertently overlooked the publishers would be glad to hear from them.

The authorised representative in the EEA is
Bonnier Books UK (Ireland) Limited.
Registered office address: Floor 3, Block 3, Miesian Plaza,
Dublin 2, D02 Y754, Ireland
compliance@bonnierbooks.ie

www.bonnierbooks.co.uk

Simon:
*To my wife Emma for all her support
and to our wonderful children Noah and Nell.*

Clare:
*To my best weather critic Sienna
and my incredible husband Chris.*

CONTENTS

INTRODUCTION 1

THE SUN 5

THE ELEMENTS 31

CLOUDS 79

THE BIGGER ATMOSPHERIC PICTURE 103

CYCLONES, HURRICANES AND TORNADOES 157

WEATHER PHENOMENA 191

WEATHER, SPACE AND
PLANETARY INFLUENCES 227

THE TECHNOLOGY OF WEATHER 251

WAR AND WEATHER 275

CLIMATE CHANGE 289

GLOSSARY 329

ACKNOWLEDGEMENTS 344

INTRODUCTION

We are all weather forecasters.

On a daily, even hourly, basis most of us assess the weather. It has been programmed into us from early humans; observing and understanding the changing sky has left an indelible mark.

Our world is a fine balance between embracing the elements and surviving them. It's only natural to look to the sky in search of patterns that will tell us what will happen next. The hues and shades, the swirls and shapes as they dance across the horizon are clues to how we will feel when we next step outside. From living off the land for thousands of years to the present-day economic and environmental impacts of climate change, the weather is an integral part of our lives. Yet, there is more to our relationship with weather than just continually adjusting to current conditions, wishing for snow at Christmas or sunshine at the weekend. Shedding light on 'the why' of the meteorological mayhem that plays out above us is a highly satisfying endeavour and the reason why many of us chose this particular practice as a career. This book is for those, like us, who are meteorologically curious.

We are passionate about meteorology and it has been a major part of our lives for decades. Simon became fascinated in the weather from the age of seven when the Great Storm of 1987 battered southern England, bringing a huge amount of damage and disruption. For Clare, understanding patterns in the atmosphere and oceans through mathematics and physics has been a lifelong journey of incredible discovery that began during her early years.

We are both qualified meteorologists with extensive training in the UK Met Office and we talk about the weather a lot. We also get asked questions about the weather a lot!

This book is a fantastic addition to any conversation about the weather. It is loaded with fascinating facts and figures, and answers many of those frequently asked meteorological questions. We illuminate common queries and unpick the more obscure and slightly surprising weather and climate conundrums.

So, let's embark on this journey together as we dive into the complex, beautiful and awe-inspiring world of weather.

Simon King and Clare Nasir
September 2019

THE SUN

WHY IS THE SKY BLUE?

On a daily basis (well, when it's not cloudy!), we accept the fact that the sky is blue. Air isn't of course blue but, simply put, light from the Sun passing through the atmosphere and into our eyes appears blue. To understand this further we need to look at how light travels through the air. While the Sun appears as a yellow/orange disc in the sky, the light originating from the Sun is actually white. White light comprises the whole spectrum of colours of the rainbow – red, orange, yellow, green, blue, indigo and violet. Each of these colours has a slightly different amount of energy in which it travels through the sky: we call this the wavelength. As the white light from the Sun passes through our atmosphere, ice, water droplets and molecules of gas scatter the light into the different colours mentioned above. This is known as Rayleigh scattering, named after the 19th-century British physicist Lord Rayleigh. Blue is scattered much more efficiently towards our eyes than all the other colours and our eyes are more attuned at detecting shorter wavelengths (where the blue colour sits on the spectrum of colours).

On a clear sunny day, the blue colour will be more vivid looking straight up and towards the Sun rather than nearer the horizon. This is because the white light from the Sun is going through less atmosphere to reach our eye and isn't getting scattered as much as it would be when we look towards the horizon, where it'll be a lighter/milkier blue.

HOW DOES THE SUN INFLUENCE EARTH'S WEATHER AND CLIMATE?

The Sun is central to the existence of Earth and the rest of our Solar System. There are a number of fundamental ways that the Sun affects Earth's weather and climate: it provides light, heat and its incredible gravitational force maintains planetary orbit. The Sun's surface is a cauldron of angry, turbulent gases exploding immense energy into the surrounding space and this energy sometimes floods into the realms of Earth's atmosphere. Before we discover the magic that the Sun's energy casts over Earth, let's get to know the vitals of our star itself.

The Sun has a radius of 695,510km and the Earth has a radius of 6,371km. To put this in perspective, 1.3 million Earths could fit into the volume of the Sun. It is widely suggested that the Sun is 4.6 billion years old, so just slightly older than Earth at 4.5 billion years. The Sun is made up of 92% hydrogen and just under 8% helium as well as other trace elements including oxygen, carbon and nitrogen, and together with intense pressure and temperature it manifests as a gigantic nuclear fusion reactor. Even though it's a nuclear-powered ball of gas it has an internal and atmospheric structure. Away from the intense heat of the core, the temperature of the Sun cools down from about 15 million °C to 2 million °C, so not enough to sustain the immense fusion process. The surface of the Sun, where the star emits its visible light, is a much cooler 5,500°C. However, the outermost layers of the Sun's atmosphere are significantly hotter yet again, with temperatures rising to 2 million °C.

> ### The Structure of the Sun
> **Core:** the gravitational attraction at the core results in immense pressure and temperatures reaching 15 million degrees °C. Nuclear fusion creates helium out of hydrogen atoms being fused together, this energy radiates outwards into the other layers of the Sun and ultimately escapes into space.
>
> **Radiative zone:** this surrounds the Sun's core and plays a key role in transferring fusion energy outwards as photons (waves or packets of light). It takes up to hundreds of thousands of years for any photon to clear the radiative layer as the process involves repeated absorption and re-emission by the Sun's gases. The layer makes up about 45% of the Sun's radius and cools with distance from the core. This zone loses about 13 million °C in heat as it connects with the outer convection zone.

WHAT DOES RAIN SMELL LIKE?

> **Convection zone:** this zone is about 2 million °C and is the outermost zone of the Sun's interior. The energy received from its neighbouring radiative zone is transported to the surface of the Sun through the process of convection (rising and falling of heat). This can be seen close to the surface of the Sun as cells of dark (falling matter) and light (rising matter). As the photons reach the surface of the Sun, light is created and this is what is seen from Earth. The radiative and convection zones are both cooler than the core.

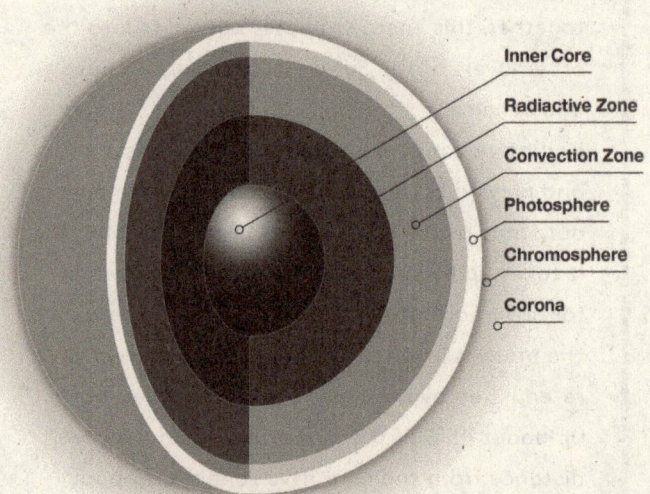

The layers of the Sun

Does the Sun Have an Atmosphere?

Like Earth, the Sun has its own atmosphere, which can be broken down into three spheres:

- **Photosphere** is the inner atmosphere, which radiates light with a temperature of roughly 5,500°C. This sphere is about 500km thick, and can be seen as spewing plasma and darker cooler sunspots.
- **Chromosphere** is hotter than the photosphere and can only be seen as a red glow during a solar eclipse. Temperatures rise to about 9,000°C at the top of this layer.
- **Corona** is the hottest layer of the atmosphere. It is 300 times hotter than the photosphere, reaching 2 million °C. While normally not visible from Earth, sometimes it can be detected during a total solar eclipse as white plumes or streaks of ionised gas that are emitted from the layer. As these gases cool they become the solar wind. There is debate as to why this outer sphere is the hottest, but some scientists have suggested explosions on the surface of the corona produce an intense energy, equivalent to a ten-megaton hydrogen bomb and millions of these go off every second in this zone.

Does the Sun Spin, Rotate, Orbit or Wobble?

It wobbles, but only small amounts due to the subtle influence of the planets' own gravitational forces on their mother star – the Sun. It spins, but not as we know it. Earth spins as a solid structured geoid with its body remaining in place, whereas the Sun is a huge ball of gas and doesn't act like a solid when it spins. There are different rates of spin and in different regions of the Sun. The Sun, along with its Solar System orbits the Milky Way Galaxy – our Solar System forms one of its arms. In turn, the Milky Way Galaxy is moving towards the Andromeda Galaxy.

How Does the Sun's Light Affect Earth?

There is a constant stream of the Sun's radiation through the year. The intensity at which it hits Earth's surface depends on what time of year it is and latitude (how far away from the equator). Before land and sea can be heated, the Sun's light has to be converted into heat or infrared energy. This transformation happens when light hits a surface. A reflective surface will re-radiate less heat than an absorbent one. A measure of how much light a surface reflects is known as its Albedo. Interestingly, very few surfaces on Earth are totally reflective (with an albedo of 1) or absorbent (an albedo of 0). For example, fresh snow has an albedo of 0.8 and a forest has an albedo of about 0.15. Cloud cover will block and reflect some sunlight, and similarly for white surfaces, like snow, the light is reflected, whereas darker surfaces, such as forests and oceans, will absorb more light.

The light that hits Earth is either reflected or absorbed, or more especially something in-between as there is rarely total reflection or absorption. The depth of influence of sunlight depends on what it falls on. If on solid ground, it doesn't penetrate very far, so it heats up this shallow layer far more than, say, a fluid layer, like the sea, where the sunlight penetrates further down, diffusing the light, which ultimately transforms to heat over a greater volume. This is why deserts have high heat during the day but cool swiftly after dark as the ground releases the heat quickly at night. The sea warms gradually over the spring and summer months and only slowly releases this heat. This has an incredible effect on moderating the air temperature, both above the water mass and of the adjoining land. Coastal districts tend to have milder winters with minima (minimum temperatures) not as low as inland, and less hot summers, where the maxima (maximum temperatures) tends to be higher inland. As the heat is re-radiated into the atmosphere it is circulated around the globe. Earth's atmosphere acts as a blanket and thus contains a good proportion of this heat. This explains why the Moon is so cold; it may have a surface for the Sun to land on but no atmosphere to trap the heat. The Sun's light not only transforms to heat but also converts to chemical energy in plants through photosynthesis, another life-giving process that sustains Earth.

How Does the Sun's Light Vary North to South Over Earth?

The position of Earth relative to the Sun means that the equator receives the most solar radiation. The Sun is directly over the equator during the spring and autumn equinox (equal day and night), so the equator gets the most direct sunlight. The Sun reaches its most northerly or southerly point during the summer and winter solstice; summer solstice being the longest day of the year and winter solstice being the shortest day of the year when the Sun has reached its most extreme point on Earth in the opposite hemisphere.

The Midnight Sun

At the poles, the maximum solar radiation happens during their summer solstice, but unlike the equator, the Sun's rays are slanted or at an angle. Around this time, there is no darkness, only light. The region north of the Arctic Circle or south of the Antarctic Circle experiences this 'Midnight Sun' in their summer. The number of days of Midnight Sun increase towards the poles. However, between 12th June and 1st July, the Arctic Circle is in constant daylight. Over the Antarctic Circle, the Midnight Sun is over two weeks either side of 21st December.

Polar Nights

Conversely, a few weeks either side of the winter solstice the sunlight disappears below the horizon completely, plunging Earth's extremities into total darkness or a long 'polar night'. This is when Earth's lowest temperatures are recorded, Antarctica holding the official record for the lowest-ever temperature, since records began, of -89.2°C. By using satellite data, scientists have discovered that in some parts of Eastern Antarctica the temperature can get below that, -98.6°C observed in July 2004. Even though the Sun disappears below the horizon during the autumn equinox, there is a spell of twilight that gets successively weaker before it finally loses all hint of light; for the North Pole this happens from mid-November, and it remains dark until the end of January. The Sun reappears during the spring equinox. In a way, for the North Pole, noon is during their summer solstice and midnight at their winter solstice.

HOW DOES EARTH HAVE ITS FOUR SEASONS?

The strength of the Sun determines the seasons. This isn't about how close Earth is to the Sun, although it is true to say that Earth's orbit is an ellipse. The reason why Earth has seasons is down to the planet's axis of rotation being tilted, currently at 23.4 degrees (although this does vary slightly through time). As Earth orbits the Sun over a year, the tilt remains in place, resulting in each hemisphere tilting away from the Sun during their winter and tilting towards the Sun during their summer. When a hemisphere is tilted towards the Sun, the Sun's rays are far more concentrated, and hence the air is warmer. And the converse is true during winter. For regions closer to the equator this means drier seasons and wetter seasons, as wind responds to the change in solar radiation and distribution of heat at surface level. For the mid-latitudes this results in a transition from autumn to a colder winter season through spring to summer. Without Earth's tilt, there would be no seasons.

WHAT IS ULTRAVIOLET LIGHT?

Ultraviolet light is part of the electromagnetic (EM) spectrum of light. The generic term 'light' describes electromagnetic energy emitted from the Sun. The light can be broken down into parts depending on how much energy they possess in terms of wavelength and frequency. The Sun emits a broad and continual range of wavelengths. These wavelengths fall into a range of categories:

- *Radio waves*: the lowest frequency and low energy. Radio waves have the longest wavelength, ranging from 1cm to 100km. Used for communication, they carry information or signals that transmit from one place to another. Radio and television stations, as well as mobile phone companies, all use radio waves to transmit signals. Stars and planets also emit radio waves. These can be received by radio telescopes on Earth that pick up the radio frequency portion of the electromagnetic spectrum.
- *Microwaves*: the second lowest frequency. Their wavelength extends 1mm to 30cm. They can penetrate objects, making fat and water vibrate to cause heat, and this is why they are used in microwave ovens. They are also used to transmit data, used in mobile phones and WiFi.
- *Infrared*: the lower to mid-range of the electromagnetic spectrum emits infrared energy, which is, in effect, invisible heat. Not all infrared energy produces heat.

In the broadest sense this part of the EM spectrum ranges from a few millimetres to 750 nanometres or 0.75 microns. These shorter wavelengths are used in imaging technologies, while the longer frequencies emit heat. Radiation is one of three ways heat is transferred around Earth (the others being convection and conduction). In this case, sunlight that impacts Earth's surface is re-emitted as infrared heat energy.

- *Visible light*: This is the light that the human eye can detect. This part of the spectrum is broken up into the colours of the rainbow. The lower frequencies emit the red light, through to the higher frequencies, which emit blue light. Objects absorb and reflect different wavelengths of light; the colour that we see is based on what is absorbed and reflected. So, a black object absorbs all colour and that's why it looks black, whereas a white object reflects all colour and that's why it looks white, and many combinations in between.
- *Ultraviolet*: This group of EM waves gets a lot of press. It's invisible to the naked eye and can't be felt, but ultraviolet (UV) radiation is responsible for tanning and burning skin with even greater exposure. However, a small level of UV provides an essential dose of vitamin D. On a broader scale, UV radiation is used for industrial and medical processes such as killing bacteria and creating fluorescent effects.
- *X-rays*: these are very high frequency waves that carry a huge amount of energy. They are emitted from the Sun's

corona. Only very hot gases emit X-rays. They don't get through Earth's atmosphere as it acts as a thick wall to these rays but they are emitted by earthly bodies. An X-ray machine fires intense beams of electrons into a small space. This produces enough energy to produce X-rays. X-rays pass easily through soft tissue but not bones, allowing for fractures to be detected.

- *Gamma rays*: the shortest wavelength, the highest frequency and, therefore, the highest energy. These rays don't get very far. As they arrive into the outer layers of the Sun, they get absorbed by plasma and re-emitted at lower frequencies. It is almost impossible to make a distinction between the highest X-rays and gamma rays, however the origin of these two sets of waves is different. Gamma rays are produced during the nuclear decay by nuclei of atoms and X-rays are produced by electrons.

All waves transport energy on Earth and in space. While air, sound and water transmit energy through mechanical waves or disturbances, they need a medium to propagate. Electromagnetic waves don't need this as they travel as waves or packets of light (photons) and can propagate through a vacuum – space. Although these EM waves exhibit different characteristics, they all travel at the same velocity at about 300,000km/s, through space. When they hit Earth's atmosphere, it all changes: only certain wavelengths are able to penetrate the atmosphere and fewer make their way to the surface. Earth's atmosphere may

THE SUN

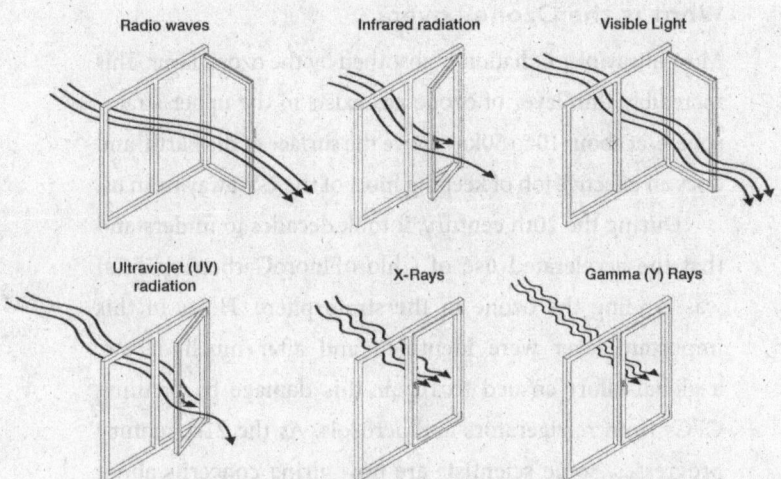

The Earth's atmosphere is transparent only to certain wavelengths of light that uses the analogy to open closed or partially-closed windows.

seem transparent, but its layers are impervious to X-rays and gamma rays, which is a good thing as they are damaging to humans.

Visible light, of course, makes it to the surface. Some radio waves flow through into Earth's atmosphere, while others don't, they bounce off the ionosphere (the layer of Earth's atmosphere above 85km which contains a high concentrations of ions and electrons that reflect some radio waves), a similar set-up for infrared and also ultraviolet radiation. Some get through, others bounce back into space and some are absorbed by the upper protective layers of the atmosphere.

What is the Ozone Layer?

Most ultraviolet radiation is absorbed by the ozone layer. This incredibly thin layer of ozone gas exists in the upper stratosphere, at about 10 to 50km above the surface of the Earth, and does an effective job of keeping most of the UV away from us.

During the 20th century, it took decades to understand that the accelerated use of ChloroFluoroCarbons (CFCs) was eroding the ozone in the stratosphere. Holes in this important layer were identified and after much debate, a global effort ensued to repair this damage by banning CFCs from refrigerators and aerosols. As the 21st century progresses, some scientists are now airing concerns about a thickening of the ozone layer having the opposite effect of reducing the amount of UV that reaches Earth.

How Do Different Types of Ultraviolet Light Affect Us?

Ultraviolet (UV) radiation can be broken down into a sub-spectrum of UVA, UVB and UVC, all of which have different wavelengths:

- UVA (315–400nm): near UV, longest wavelength and penetrates the atmosphere;
- UVB (280–315nm): middle UV, 90% absorbed by the ozone layer, but 10% penetrates the atmosphere;
- UVC (100–280nm): far UV, shortest wavelength, all absorbed by the ozone layer and doesn't reach Earth's surface.

UVA is the most common radiation to affect life on Earth (about 95% of all UV). It is also the tanning ray that is used in tanning machines. Exposure to too much UVA is known to cause skin cancer. Although UVA is able to dive deep into the skin layers (dermis and subcutaneous), UVB fails to get any deeper than the surface layer of the skin (epidermis). Still, exposure to UVB is responsible for burning and the redness of the outer layer of the skin. It also plays a key role in skin cancer and is most prevalent during the middle hours of the day.

What is the Ultraviolet Index?

We cannot see or feel UV rays on our skin but they are a major cause of skin cancer around the world. When the weather report highlights a high UV forecast extra precautions need to be taken to protect our skin from the Sun's harmful rays.

The level of UV radiation reaching the surface is calculated using computer models that consider not only the intensity of solar radiation at ground level, which includes cloud cover, but also the thickness of Stratospheric ozone, the elevation of the ground from sea level and other factors such as time of day and the gas composition of the lower layers of the atmosphere. In 1992, the World Health and World Meteorological Organizations of the United Nations introduced a scale that is directly proportional to the intensity. The lowest level is 1 (no/very low radiation – essentially at night) up to the highest: 11+ (extreme risk or harm from UV radiation).

How Does Ultraviolet Light Vary?

The amount of UV radiation reaching Earth's surface on any given day or location varies throughout the year in relation to the position of the Sun in the sky. Over the equator, the Sun is high in the sky for most of the year and so the level of UV remains consistently high, but the further away from the equator, the high levels of UV become more of a seasonal feature. During the winter months, the Sun is low in the sky so the amount of UV radiation reaching the surface is lower than during the height of the summer months when the Sun is higher in the sky and the UV radiation has a more direct route through the atmosphere to Earth's surface (importantly, passing through less ozone), therefore keeping more of its energy.

For mid-latitude countries, such as the UK, the highest UV levels occur during June when the Sun is highest in the sky. The highest level of UV in the UK during this time is normally around an index level of 7 or on rarer occasions 8. UV levels in April are very similar to those in August and mid to late spring is generally cooler than the summer months. So, in April, people tend to be more vulnerable to sunburn than, say, August; the false perception is that the Sun isn't as strong – it may not feel as warm, but UV levels are in fact very similar. The time of day is also an important consideration as between the hours of 10am and 3pm local, the Sun gets high enough in the sky for the UV level to start to creep up.

Other factors that can determine the UV level for any location are the cloud cover, altitude, land cover and atmospheric ozone.

- *Cloud cover*: on very cloudy days the UV level can be reduced (on some cloudy days, the clouds can be thin and you can still get high UV levels).
- *Altitude*: for every one thousand feet, or 300 metres up a mountain, the level of UV will increase by 2% due to the thinning of the air.
- *Land cover*: different land surfaces reflect UV radiation. At the beach, seawater will reflect up to 10% more UV and the sand as much as 15% than darker surfaces such as grass or generally inland of coastal strips. Similarly with ice and snow, the whiteness of the surface will reflect as much as 80% more UV. This all leads to an increased risk of being sunburnt.
- *Atmospheric ozone*: ozone in the atmosphere filters UV radiation so monitoring the thickness of this layer is vital in determining the intensity of UV that will ultimately reach the ground. The amount of ozone above our heads varies latitudinally, longitudinally and seasonally. Sometimes the layer thins significantly in one part of the sky. Holes in the layer can also form during particular times of the year, for example the Antarctic spring time.

HOW DOES EARTH'S ROTATION AFFECT WEATHER PATTERNS?

Everything in space rotates, from asteroids and planets to solar systems and galaxies. This is known as the conservation of angular momentum and is the legacy of the early formation of our universe when gas and dust collapsed forming the Sun and the planets around it.

Once things started to move in space, they kept on moving. This is called inertia. Earth not only orbits the Sun, it rotates or spins on its own axis. This spin is essential for the order of things on Earth as without spinning Earth would exist under a new order. By losing its centrifugal force, gravity would dominate, oceans would migrate to the poles where gravity is strongest, which would lead to a revealing mass of lands traversing the equator to the mid-latitudes. An Earth-year would remain the same amount of time, but a day on Earth would last the same as a year. These two factors alone would completely overhaul Earth's climate; catastrophically for life as we know it today. Earth's spin creates balance and movement across the oceans and the atmosphere, and because of this spin certain forces are present and remain in play.

The Coriolis force is a frictional force that deflects fluids on a rotating frame of reference (Earth). It acts perpendicular to the axis of rotation and is known as an apparent force because it's experienced from the point of view of an observer on Earth, rather than the actually moving body of fluid.

The surface of Earth moves at a different rate to the air above it. What is a deemed a straight path of a passage of air aloft will be seen to be deflected at surface level, as the Earth continues rotating. In the Northern Hemisphere, this deflection is to the right of the path, or has a westerly component. The Coriolis force is greatest at the poles: the closer to the equator, the less this force makes its mark on the weather.

The Coriolis force is instrumental in creating the spirals of cloud and air that transport moisture and heat across the planet. Mid-latitude low pressure systems with their notorious dartboard pressure pattern are due to forces related to the Earth's rotation when warm air moving north clashes with cold air spreading south. The extra deflection can grow into a circulation as these two air masses entwine. On satellite imagery the magnificent swirls of cyclones, with their defined eye in the centre of the storm and flailing clouds that spin outwards, are stunning evidence of this apparent force. The infamous trade winds that exist at 30 degrees north and south are another example of how the Coriolis force shapes global weather patterns. North-easterly winds, north of the equator converge with south-easterly trade winds south of the equator. Both are deflected from a straight northerly or southerly path by the Coriolis force (to the right in the Northern Hemisphere, to the left in the Southern Hemisphere). The region where these winds meet is called the Intertropical Convergence Zone (ITCZ) and is responsible for the low pressure belt that circles Earth's atmosphere close to the equator.

WHAT ARE SOLAR STORMS AND HOW DO THEY AFFECT EARTH?

Although from Earth the Sun seems like a distant ball of fire, its influence is more than light and gravitational pull. The surface of the Sun is energised and active, with bursts of highly charged particles continually exploding into space, and Earth is sometimes in the firing line.

The Solar Wind

There is a continual flow of highly charged particles that are emitted from the Sun: this is the solar wind. During intense explosions of plasma, known as Coronal Mass Ejections (CMEs), this steady flow of solar wind is disturbed. The plasma explodes out in all directions and accelerates to speeds as high as 3,000km/s and temperatures of 1 million °C. The most incredible property of this plasma is that it has a magnetic field and is attracted to Earth's own magnetic field, the strongest attraction being at the North and South Poles. Occasionally, these highly charged particles stream towards Earth, enveloping the planet with the most intense concentration towards the poles. The greens, blues, reds sometimes form hazy curtains of light, other times a defined spectrum of swirls are all evidence that the Sun's bodily gases are closer than 150 million km away. The solar wind, attracted by Earth's own magnetic field, is sucked into and then clashes with gases in the outer realms of Earth's atmosphere, namely the thermosphere. Collisions of the Sun's charged particles with oxygen produce red and green auroras, whereas nitrogen creates the purples and pinks. This breathtaking drama of dancing light that plays out at the poles through the year is something on many people's bucket lists. These spectacular light displays can only be seen during the dark and coldest months; through the summer, the Sun hardly leaves the sky close to the Arctic Circle.

Solar Flares and Coronal Mass Ejections (CMEs)

Solar flares are large eruptions of electromagnetic radiation from the Sun's surface. They propel highly charged solar particles into space, but there are more violent events that can have a profound effect on Earth. These are called Coronal Mass Ejections, which are a more immense and powerful version of solar flares. It takes three to four days for these solar storms to reach Earth, enough time to anticipate what damage could potentially be caused. The energy in these particles penetrates through the outer layers of the atmosphere where the auroras form. They affect Earth's magnetic field and produce many problems with power grids, radio communications and satellites.

The Carrington Event

In 1859, a few days prior to 1st September, a Coronal Mass Ejection leapt from the Sun's corona and flew through space towards Earth on an intensely strong solar wind. There were auroras across the Southern and Northern Hemispheres. This was very unusual, as normally auroras tend to appear close to the poles. During this event they extended towards the equator as far north as Queensland and as far south as Cuba. However, this electromagnetic storm did more than produce incredible visuals across the upper echelons of Earth's atmosphere. Telegraph networks across many parts of North America and Europe failed as sparks flew from pylons.

Although such events are rare, the Carrington Event is not unique in history. Even as recently as 2012, a similar Coronal Mass Ejection narrowly missed Earth. Such an intense solar wind would have had far more costly and extensive consequences in this digital age.

Two decades earlier, a Coronal Mass Ejection with the power of 20 million nuclear bombs shot a cloud of protons and electrons into space. Carried by the solar wind, this hit Earth and resulted in cutting out the Hydro-Quebec power grid – 6 million people lost power for nine hours. Although a smaller event than Carrington, it's a tell-tale sign that these solar storms are a reality, with lead times of days rather than weeks.

SUNSPOTS AND SOLAR CYCLES

Each solar cycle lasts about 11 years, peaking at a solar maximum, and eventually dipping to a solar minimum. Each cycle relates to the strength of the Sun's magnetic field – the strongest field peaks at the solar maxima, and this is where the most sunspots occur. During a solar minimum, it's rare to observe sunspots and therefore sunspot activity is a good indicator and measure of the Sun's solar activity and in turn, this may indicate how Earth responds.

> ### Sunspots: what are they?
> **Colour:** sunspots are dark blobs or patches that appear on the surface of the Sun. They have two parts: the umbra (the darker part) and the penumbra (more shadow than dark).
> **Position:** on more careful examination sunspots are located on the photosphere, which is the inner layer of the Sun's atmosphere and generally cooler than the outer layer (the chromosphere and corona);
> **Temperature:** sunspots are cooler still than their surrounding surface environment at around 3,700°C, so at least 1,000 degrees lower than the rest of photosphere;
> **Size:** sunspots can be many times the size of Earth;
> **Formation:** sunspots form due to the Sun's internal magnetic field that winds its way to the surface, where they form a sunspot;
> **Location:** sunspots tend to appear around distinct regions of the Sun's sphere, 15 to 20 degrees either side of the Sun's equator, and are never found north or south of 70 degrees;
> **When:** sunspots tend to appear when the Sun is in its solar maximum, which means when the Sun is at its most active.

Sunspots are a reliable sign that the Sun's solar activity is strong, as the plasma explosions occur close to sunspots and this results in solar flares and the bigger beast, Coronal Mass Ejections. The solar wind then carries this mass of highly charged particles into space, where sometimes it surges in the direction of Earth, attracted by its magnetic field. During 1645 to 1715 there was a solar minimum, where sunspot activity was near to zero. This coincided with the Little Ice Age. The period was coined the Maunder Minimum, yet still heated debate rages amongst some scientists today as to why it was so cold. Was it a consequence of the solar minimum? It's worth noting the Little Ice Age also happened around the same time as a rise in volcanic activity that spewed more particles into the upper atmosphere, blocking out the Sun's rays and reducing the effect of the Sun further.

The bigger question – is there any evidence that these solar changes affect the climate? Well, in some ways yes, as the Sun's energy ebbs and flows over time, decades and centuries, the solar maximum is associated with higher UV rates, which affects life on Earth and the workings of the atmosphere. During solar minima, when the sunspot activity is at its lowest, UV radiation is lower than at the maximum.

This of course has an effect, but it must be viewed against the background of increasing pollution and in particular, greenhouse gases, alongside Earth's response to this, so it's almost impossible to delineate the solar activity from the climate change and from the natural fluctuation of Earth's climate.

THE ELEMENTS

The Sun controls Earth's weather on every level. One hundred and fifty million kilometres away, the Sun's influence cascades through every layer of atmosphere, determining changes in sunlight, temperature, humidity and atmospheric pressure. These ever-changing properties of air, in every permutation and combination, create the weather elements. In their most basic forms these are wind, cloud and precipitation. This masks a family of far more specific elements, in every shape and form; snow, fog and jet streams among many others. By breaking them down, we can see exactly how they affect the world around us.

WIND: THE MOST INFLUENTIAL WEATHER ELEMENT ON EARTH

Wind is probably the most significant weather element on Earth. We often pass it off as one of those annoying things that can mess with our hair, drive rain into our faces or bring severe destruction in the form of hurricanes or tornadoes. But, the wind is incredibly important on both a global and local scale. Globally, we need wind to transport warm air from the equator to the poles to regulate the temperature. We need wind to distribute moisture around the world – the hydrological (water) cycle wouldn't exist without it. On a local scale, wind is one of the main ingredients in pretty much every weather we experience. And it's not just in meteorology where wind is really important, our geography and natural landscapes are shaped by the wind too. From erosion of rocks to the dispersion of pollen from plants and flowers.

Starting with the basics, the wind is very simply defined in meteorology as the flow of air. This flow of air can either be vertically, via up- and downdrafts, or horizontally, from any direction. In a regular weather forecast you will be updated about the wind speed and direction. The latter is always cited as where the wind is coming from, i.e. a south-westerly, which means the wind is coming from the southwest. Wind speeds can vary from the typical calm 0–10mph up to hurricane force speeds of over 74mph, all the way up to tornadic and high-level jet stream strength, with winds up to 300mph.

To understand how the air moves, temperature and pressure need to be considered. When the Sun heats air, atoms and molecules become excited and the space between them grows as they vibrate and move more frantically. This warming air thus expands and begins to rise. Conversely, in cooler air, the atoms and molecules stay closer together, making it denser and therefore causing the air to sink. This rising and sinking is how we understand atmospheric pressure; warm air rising will make the pressure at the surface lower, cold air sinking down on to Earth's surface makes the pressure higher. When there is a difference in air pressure within a part of the atmosphere, air will naturally flow from the higher pressure to where there is less air – low pressure. This movement of air is wind. The larger the pressure difference is between two areas of air, the stronger the wind. Low pressure zones produce wet and windy weather – the wind flows rapidly towards the centre of the low causing strong winds, whilst rising, forming cloud and rain.

Of course, it isn't quite as simple as that. On the global scale we know that the Sun heats the equator more than the poles. This means that at the equator, the air is being warmed and, as per the above, it expands and rises. When the air reaches the top of our atmosphere and can't go any higher, it will either be pushed north or south towards the poles. As this happens, the air will cool and then sink back down to the surface. Once it reaches the surface, it will then be pushed either north or south again. The air travelling back to the equator will eventually get warmer and start the whole process again. If you were following that you'll know that this process has generated a circulation. This is called the Hadley cell circulation and there's another at the mid-latitudes called the Ferrel cell, and one up to the poles called the Polar cell. It's these circulation cells that stop the equator from getting too hot and the poles too cold.

Let's add another dimension to this. Earth is also spinning, one rotation in a 24-hour period. This rotation creates the Coriolis force, which will make the winds bend to the left or right depending on which hemisphere you're in as they travel north and south. At the surface, the combination of the global circulation cells and the Coriolis force gives us the 'trade winds'. From the equator and tropics in the Northern Hemisphere, the trade winds flow from east to west (the reason why hurricanes originate from the west of Africa and head west to the Caribbean and USA). Over the mid-latitudes in the Northern Hemisphere the trade winds are westerly, flowing across the North Atlantic towards Europe. This dominant wind direction can sometimes steer ex-hurricanes from the USA eastern seaboard towards the east. It is also the reason why the UK's prevailing weather direction is from the southwest.

Whenever you hear weather forecasters, or indeed look at a weather map yourself, you can determine which direction the wind is blowing by looking at the position of the high and low pressures. We've explained that wind blows from high to low but as you've also got the Coriolis force in play, the wind doesn't just go in a straight line but will rotate around these areas of high and low pressure: clockwise around a high and anti-clockwise around a low.

The Family of Jet Winds

While at the surface where the circulation cells meet we get the trade winds, at the top of the atmosphere is where you'll find another type of wind – the jet stream. It is essentially a narrow band of strong winds that can start, stop or split as it meanders around the world. There are two main jet streams: the Subtropical Jet, which resides in between the Hadley and Ferrel cells, and the Polar Jet, which sits between the Ferrel and Polar cells. The movement of the latter can be quite erratic, but it travels from west to east either being zonal (straight from west to east) or meridional (taking more of a north–south movement). The speed of the jet streams can vary but typically, the air travels between 60 and 250mph. The height of a jet stream exists at around 7 to 12km (about 20,000 to 40,000ft) in the atmosphere and has a width of only 100km.

Meteorologists are particularly interested in the jet stream because it marks the boundary between warm and cold air, and its change in speed and direction influences how weather systems develop at the surface. The Polar Jet is the one most important to those living in mid-latitudes. It is driven by the temperature difference between the cold air over the poles and the warmer air in the subtropics, and is most pronounced during winter across North America, the Atlantic and Northern Europe. As the Polar Jet meanders and changes shape it can bring the UK variable conditions day to day, week to week. As forecasters, it is one of the first elements we analyse. If the Polar Jet is positioned to the north of the UK, the air tends to be warmer than average as it originates from the subtropics. If it is to the south of the UK, the air can be colder than average as the winds will have a northerly component, coming from the Arctic. When the jet stream is close to the UK and Northwest Europe, low pressure systems will track close by with the greatest chance of wet and windy weather.

The impact of the Subtropical Jet isn't as obvious as the Polar Jet, as the temperature difference between the subtropics and tropics isn't as great. The Subtropical Jet also varies much more throughout the year, becoming stronger in the winter and almost non-existent in the summer months. While it is not as influential on the weather as the Polar Jet, it is connected to the Indian Monsoon.

One other jet worthy of mention is the African Easterly Jet (AEJ). This is slightly different to the two main jet streams as it is lower down in the atmosphere at approximately 3km (10,000ft). Unlike the Polar and Subtropical Jet, the AEJ is an easterly wind originating from East Africa across the African continent to the Atlantic. It is formed by the temperature difference between the hot Sahara Desert to the north and the cooler waters of the Gulf of Guinea. It is at its strongest from late spring to early autumn, where it sits approximately from Ethiopia to The Gambia. While the maximum winds in the AEJ are only up to around 25–30mph, it is instrumental in the development of tropical storms and hurricanes in the Atlantic.

Local Winds

Mountains can play an interesting part in wind patterns. Such large features will divert winds either around them or most often over the top. When this happens, air is forced to go where it naturally doesn't want to be and will undergo changes in its pressure, humidity, temperature and sometimes strength.

One of the few officially named local winds in the UK is the Helm Wind, which blows over the Pennines in northern England. When air is stable it doesn't have much vertical motion. This is because a layer of air above it has capped the wind from going any higher, acting like a lid. If the cap is above a mountain peak up to about 300 metres (1,000ft), as the wind rises up the side of a mountain, it gets squeezed by the cap above. This squeezing accelerates the air so that when it descends down the other side of the mountain the winds will be stronger. In Cumbria, when there is a northeasterly wind, this effect can occur over the highest part of the Pennines called Cross Fell, bringing a strong Helm Wind down the southwestern slopes. As this air comes over the top of Cross Fell, it will also create an undulating pattern in the airstream and where the air crests, you can end up with cloud forming. This cloud is known locally as the Helm Bar and is a stationary, dark and menacing-looking cloud sitting downslope from the peak of Cross Fell.

Foehn Effect

The Foehn Effect isn't technically a wind but it can bring very different weather conditions to either side of a mountain range. This effect can be found all over the world but in the UK, it is most notable over the Scottish Highlands and Welsh mountains. When there's a weather system heading in from the Atlantic with an abundance of warm moist air it will meet the mountainous terrain. The air temperature may be around 10°C at this point. The air is forced to rise up the mountainside, where it will cool, condense into thick cloud and may even rain. As the air gets to the top of the mountain and comes back down the other side, it has been modified in such a way that it is drier and warmer. This change happens because saturated air behaves differently to dry air when it rises and falls in the atmosphere. That's why, to the lee of the mountain, you'll find sunnier skies, with the temperature as high as 18°C, while those on the other side of the mountain will experience very different weather.

The Mistral

In southern France and the Mediterranean, you can sometimes get intense cold north or north-westerly winds blowing, known as the Mistral. These are most common during the spring or winter but can arrive at any time of the year, with sustained wind speeds of up to 60mph that last over a week. On some occasions, the Mistral can bring violent winds up to 110mph and cause a great deal of damage. These winds occur when you get an area of high pressure sitting in the Bay of Biscay and a low-pressure system around Corsica or Sardinia. With this pressure set-up, the wind will be a north or north-westerly, which accelerates through the Rhône Valley in France down to the Mediterranean coast. The effects can be more far-reaching though than just the Rhône Valley, with Provence, Languedoc and even Corsica and Sardinia being impacted by the winds. When the Mistral blows it is often after the passage of a cold front and in this case it will bring clearer, fresher weather to the region, although not always. When it does, visibility can dramatically improve so the sky will appear even more azure and the Alps can be seen more than 90 miles away.

WATER

About 71% of Earth's surface is covered by water, amounting to about 1,386 million km^3. This water is what makes the planet look blue from space. In fact, the ocean is made up of water molecules and they absorb certain wavelengths of light more efficiently than others. The longer wavelength red light is absorbed close to the water's surface. The deeper the water, the more other colours – orange, yellows, greens – get absorbed, leaving the blue. Incredibly, 96.5% of all of Earth's water is held in the oceans. This water is saline; the rest, just over 3%, is freshwater. Freshwater comes in many guises: water vapour (gas), water as a fluid that flows into streams, rivers and lakes, and seeps into the ground to be absorbed by the soils or where it is stored in underwater lakes and aquifers. Water is also found in all life, from humans to plant life.

Water also takes solid form as ice. Ice, either as glaciers and snow or just plain ice, holds the most freshwater, about 68%. Of this, Antarctica houses 90% of Earth's ice, about 30 million km^3. To put this into perspective, if all the ice over this continent melted, sea levels would rise across the world by 58m. The other major ice sheet is over Greenland, and together with Antarctica, about 99% of total ice is locked up in these two regions. These ice sheets develop when snow falls in winter and remains in solid form during the summer months, then layers of snow pile up and grow denser, compressing the older layers below.

All water, in every form, moves. As a gas, water vapour travels at great speed, carried along by Earth's winds. Water flows to the lowest point, replenished from above. Ice sheets are also constantly in motion, flowing downhill under their own weight, eventually reaching the sea. As long as this melted ice is replenished by snow, the system remains in balance. This movement of water through its different states, continually transferring between ocean and air, is called the water or hydrological cycle. The processes recycle and replenish the quality of fresh water through evaporation, transpiration, condensation, precipitation, percolation and runoff:

- *Evaporation*: when water in liquid form becomes a gas or water vapour, this happens when a puddle dries up.
- *Transpiration*: this is evaporation from leaves of plants, instead of bodies of water.
- *Condensation*: when water changes from gas to liquid, this is seen every day as clouds materialise across the sky.
- *Precipitation*: condensation leads to precipitation, as the weight of rain, snow, hail or drizzle results in this water in liquid or solid form falling to Earth.
- *Percolation*: water also seeps into the ground, a process called percolation, and here, it is absorbed by the soils,

or it seeps into the rocks and sometimes then arrives in underwater lakes and aquifers.
- *Runoff*: water that travels across the surface of Earth, gathering or joining as bodies of water in lakes and rivers, and eventually flowing into the sea.

What's the Difference Between Rain and Showers?

On a very basic level anything that comes from the sky, that is water, is rain. We could stop there, but there is good reason why we differentiate between types of precipitation: in this case, namely rain or showers. The fundamental difference is in the types of cloud that produce the rain. Look up to the sky and there is a whole family of clouds. There are about nine distinct cloud types, based on shape, structure and height but they are generally not all present at the same time.

There are two basic types of cloud that reveal reams about their inception. These are layered clouds (stratiform) and heaped clouds (cumuli). Layered clouds form when two huge air masses, one cooler than the other, merge and condense. Compared to its 'heaped' cousin, this extended layer of cloud doesn't boast much depth. The formation of stratiform cloud is all about the movement of air horizontally, known as the process of advection, and this creates a vista of vast spreading cloud that may undulate but doesn't rise to the heavens.

Layer cloud tends to cover a lot of sky and extends for tens, if not, hundreds of kilometres in every direction. When this cloud is loaded with ever-growing cloud droplets, it eventually rains. And yes, this is actually meteorological rain in every way. The sky will appear laden with darkening clouds, producing rain for quite some time, with no hint of sunshine. Sometimes they will bring heavy rain, sometimes intermittent, and the passage of the cloud will eventually be marked by drier weather and some brighter skies.

Shower clouds form in a very different way. They are called cumulus clouds and can grow into a far more powerful and decidedly larger version – the cumulonimbus. These are huge majestic beasts that explode into the sky, extending vertically towards the troposphere's invisible ceiling. Here, the cloud spreads out like a giant anvil looming across the top of the sky.

Cumulus clouds form through vertical motions in the air. These are thermals; they rise due to the local heating of the land, rather than the en masse ensemble of warm and cold that is the signature of the layer clouds. This is the process of convection. The air is able to rise because it's warmer, thus lighter, than the environment; warm air can rise into cooler conditions. The converse is therefore also true: cool air cannot rise into warmer air.

There is a good reason why forecasters distinguish between rain and showers: it's down to the individual experience of wet weather. Rain tends to ease in, become intense and then ease out, not always, but mostly. Clouds are dominant and block out the Sun. A 'rain' shower is a shorter burst of wet weather, where winds pick up and turn quite squally and the temperature drops temporarily. Either side, there's sunshine. So, one is prolonged, the other short and sharp. Sometimes showers follow rain, as in the classical 'Norwegian' frontal model; layer cloud (rain) arrives first and gives way to heaped clouds (showers) that take up the rear. From thick cloud where no sky is exposed to open cell convection, where each individual cloud is separated from its neighbours, thus exposing the Sun from time as they march across the sky – and this is where the phrase 'sunshine and showers' comes from.

When Do We See Most Showers and Most Rain?

The distribution of cumulus clouds varies during the year. During the winter months, showers form over sea and tend to just affect coastal districts. This is because the sea retains more heat than the land. The combination of relatively warm sea and a cold source of air flow (such as from the Arctic) results in local pockets of warmth from the sea rising into the cooler upper air environment. There comes a point, about a month to six weeks after the winter solstice, when the Sun's rays gain strength and the first fledgling cumulus clouds start to appear over the land; in the UK, this can be during the month of February and is a sure sign that spring is just around the corner.

By April, stronger solar radiation has really got to work and coupled with still very cold air aloft, causes large cumulus clouds to form, resulting in heavy, and at times frequent, April showers. In fact, showers are a generic term for anything that falls from cumulus-type cloud. When the air is very cold, showers can be of sleet or snow. When the air is loaded with energy, cumulonimbus clouds dominate the horizon, hail showers will fall with the higher likelihood of a thunderstorm – hail, thunder, lightning and squally winds. And, of course, with April showers comes the promise of rainbows ...

Bands of rain from layer clouds are far more associated with low-pressure systems that develop initially over the ocean and drive across the sky helped by the jet stream. Although it can be wet during the mid-latitude summer, the temperature contrast between Arctic and tropical is far more pronounced away from the summer months. This leads not only to a stronger jet stream, but also the position of these strong upper winds tends to be halfway between Iceland and the Azores or the mid-latitudes. This is one reason why there is an increased storminess during autumn and winter months: a powerful jet stream spawning low-pressure systems over the mid-Atlantic and releasing its load of rain and wind across the UK and Northwest Europe.

WHAT IS THE SHAPE OF A RAINDROP?

The traditional image of a raindrop is a tear that droops from a pinpoint to a bulbous underbelly. Does this classical shape actually exist in the atmosphere? To understand the ultimate morphing of a drop of rain as it falls from the sky, it's important to rewind to the drop's early development.

A cloud forms from a gathering of tiny water particles as water vapour cools and transforms from a gas to a liquid clinging to microscopic dust particles (condensation nuclei). Cloud droplets are ten to 1,000 times smaller than raindrops. The average size of a cloud droplet is about ten to 15 microns (a micron is 1/1000mm). The shape of a cloud droplet is spherical. Its lack of weight means that it is easily held in suspension within the cloud in an environment of buoyant air. As the cloud becomes crowded with these spherical cloud droplets, they merge with others. A typical raindrop in its infancy is a coalescence of 10,000 cloud droplets. At this stage they are still spherical but as the raindrop gathers more cloud droplet minions, and becomes heavier, a battle commences between two forces: surface tension that acts to hold the water bundle together and air flow that pushes up against it as it falls through the sky. The larger the drop, the more pressure on the underside due to wind resistance, and it is this upward force that flattens the drop out. The differential pressure results in a curved top and flat bottom more like a squashed muffin. The greater the volume, the faster the speed of the drop and the greater the air resistance. Eventually, rather than the classical teardrop shape being realised, the air rips the drop apart and becomes a series of smaller droplets, most of which resume a spherical shape before they grow again or impact Earth.

Why Doesn't It Rain All At Once From a Cloud?

An average cumulus cloud weighs about half a million kilograms or 500 tonnes – that's the equivalent of about 100 elephants. It's hard to imagine how such a hefty beast can remain suspended in the air. However, a cloud is made up not only of cloud droplets but also a lot of air (a mix of gases, including water vapour). These tiny cloud droplets have a terminal velocity of around 10m per hour. To put this in perspective, falling from, say, 2,000m, it would take 200 hours for a typical cloud droplet to land on the ground. Air resistance and air currents, such as updrafts, are far more powerful. Clouds tend to form, survive and grow in air that is moving upwards. It's only when the droplets coalesce to form raindrops, and become 300 times larger, that the terminal velocity, or downward speed, fall to minutes.

Back to the cloud ... As we have discovered, clouds are a gathering of tiny water or ice particles called cloud droplets. There are trillions upon trillions of them in a cloud, and the darker the cloud, the more huddled together they are. With the help of winds within the cloud they collide or merge to form larger cloud droplets – the process of coalescence. A cloud dense with cloud droplets tends to be a greyer colour. When enough of these tiny cloud droplets merge a rain droplet forms, which is then heavy enough to fall to the earth. Rain droplets don't all form and fall together at one time, that's why a shower can vary in intensity. Until the air dries out, this process of cloud droplets appearing from water vapour and condensation nuclei and then growing to raindrop size is a continual one. Raindrops grow at different rates and fall only when they are heavy enough to fight against the upward forces of air. Of course, gravity will help along the way as well. So, it keeps on raining, but because every raindrop has its own development time frame, all drops in a cloud don't fall at the same time. And if there continues to be a feed of moist air into the cloud, the process of rain droplet formation carries on and it keeps on raining.

What Does Rain Smell Like?

The scent of rain has been around since the first of the continents rose up and separated from the oceans, as storm clouds parted allowing sunlight to energise the air, warm the seas and dry the newly exposed lands. But it wasn't until 1964, when the smell of rain finally received its name – Petrichor. Petrichor describes the scent of the air just before, during and after it rains, more especially after a dry spell. It is a modern-day description of an ancient and life-affirming smell. 'Petra' translates to stone and 'Ichor' to 'blood from the veins of heavenly bodies', with the literal translation of Petrichor being 'heavenly fluids from a stone'.

The first scientific study on Petrichor was carried out by two Australian scientists in the 1960s, who analysed every part of the process that makes up the scent. Before the rain, increased moisture loosens earthy compounds, allowing gases and particles from the dry ground to diffuse into the atmosphere. As it rains, the action of raindrops hitting the dry ground releases gases from the pores of stones and rocks – essentially, particles from the ground fly up into the air. The scent becomes more intense. After it rains, the smell of rain lingers in the moisture-laden air. The process is part mechanical, the action of raindrops, and part chemical, as liberated bubbles of gas and particles bounce and burst into the air.

Petrichor is a scent that generations have become familiar with, particularly in parts of the world where the seasons divide into wet and dry. The anticipated rainy seasons brings the promise of water, crops and relief from a hot, dusty climate. This anticipation is all-too-familiar every year across India and its subcontinent. Pre-monsoon heat intensifies during March, April and May. Burning sunshine pushes temperatures into the 40°C+ and even 50°C+. There is no escape from the stifling, stinging and dirty air. Imagine the immense relief of millions when the monsoon rain arrives: the air is cleansed and the parched land quenched. These heady atmospheric aromas transform into a fresh scent that is celebrated across the breadth of the continent. Petrichor – the smell that guarantees rivers will again flow and crops will thrive for another year. The smell is so revered that a small perfumery industry in the village of Kannauj, nestled in the province of Uttar Pradesh, is dedicated to bottling this perfume from the heavens. For hundreds of years, generations have trapped and produced these earthy fragrances. The oils they use are called attars, formed from dry discs of soil that are infused by moist air and water droplets, allowing this subtle fragrance to be tapped. After all, it's the smell of India, it's the essence of life.

A more familiar scent across the western world is Geosmin, the name given to the smell made by the impact of raindrops on tarmac and soils – a mix of organic materials, earthy bacteria and ozone.

What Exactly Is 'Wintry Precipitation'?

During the winter or even spring in the UK, you may hear a forecaster saying 'wintry precipitation' and you may wonder what on earth they're talking about. Well, precipitation is the collective term for any weather element falling out of the sky, whether it is rain, sleet, hail or snow. While for many, the default precipitation is rain, when the colder months come, we can increasingly see this precipitation turn into snow, sleet and freezing rain.

How Does Hail Form?

Hail can occur any time of the year but in the UK it is most frequent during the winter and early spring. However, some of the most dramatic hail can fall during the late spring and summer when there's enough energy in the atmosphere for cumulonimbus or cumulus congestus clouds to develop. These are the clouds that stretch high into the atmosphere where the air is below freezing. Within these clouds, supercooled water droplets are thrown up and down by the up- and downdrafts. As these water droplets are thrusted into the upper part of the cloud, they freeze into a tiny ice pellets of hail. The process continues as they then descend, falling into air that is above freezing and abundant in water droplets. This coats the ice pellets with a thin layer of water before updrafts yet again push them back into the sub-zero zone of the cloud – where they freeze again. This cycle of layering will continue many times until the hail is heavy enough to counteract the updrafts. At this point, they will fall out of the bottom of the cloud to the ground. If you were to cut a hailstone in half, you'd be able to see the concentric layers made during this process, just like tree rings.

The size of the hail will depend on how many cycles of refreezing the hailstone goes through, which is linked to how strong the up- and downdrafts are within the cloud. If you've got a big summertime cumulonimbus cloud that has been formed on a very warm day, the updrafts are going to be strong. In this case, the hailstones will remain in the cloud for some time and undergo numerous refreezing cycles. This results in large hail. Some of the largest hail can grow to the size of golf balls before making a surprise impact. The largest hail stone ever to be recorded had a diameter of 20.3cm. When it fell to the ground in South Dakota, during the summer of 2010, it weighed 0.88kg! Hail this huge can cause serious injury and damage. In fact, hail is probably one of the most underrated weather elements in terms of damage potential. During big hailstorms in the USA, cars, crops and buildings become damaged, with insurance costs running into millions of dollars.

One example of how damaging hail can be is from April 2013 when, at Kandahar airport, Afghanistan, a 30-minute hailstorm containing hail that was about the size of golf balls hit the airfield. At the military base, the hail smashed up car windscreens and anything left outside, including British and American aircraft. The damage caused cost the British government £13 million in repairs and write-offs. It also meant there was a significant reduction in military capability while aircraft were unusable.

Freezing Rain: That's Just Ice, Right?

You may not have heard about freezing rain as it is quite a rare weather phenomenon, but it can be one of the most hazardous types of winter weather. During the winter, most precipitation starts out as either supercooled liquid water droplets, snow or tiny ice pellets in the cloud. On occasion, when the lower part of the atmosphere known as the boundary layer is cold (below roughly 2°C), any precipitation will fall as sleet or snow. In some weather situations though we can get a thin layer of warm air close to the surface. As the precipitation passes through this 'warm nose', it will warm slightly and turn into rain, but as it passes out of this warm layer back into the cold air nearer the surface, it will cool enough to form supercooled liquid water droplets.

Supercooled water is when water droplets can still exist as liquid, even when the temperature is below freezing. As these supercooled water droplets continue to fall, they will eventually hit the ground and freeze instantly. Freezing rain can coat objects with a layer of ice. Cars, roads, power lines and rail lines will become very icy, causing significant hazards and disruption. While the UK doesn't get many freezing rain events, in other countries such as the USA there can be major freezing rain storms that create widespread disruption. Freezing rain can turn roads into ice rinks, rendering them unusable, and even trees and power lines can be brought down by the weight of the ice building up on them.

Sleet

This is probably the easiest type of wintry precipitation to explain as it isn't really a unique type of precipitation. In fact, in meteorology there is no such thing as sleet. In an official weather observation, we would report what you know as sleet actually as 'rain and snow mixture'. If snow and ice pellets originating from a cloud fall into air near the surface that is slightly warmer (around 0–2°C), the snow and ice will start to melt back into liquid water droplets – rain. If the warmer layer of air isn't particularly thick when the melting process starts, it may not have fully finished before it reaches us, leading to a mixture of partially melted snow, ice pellets and raindrops reaching the surface.

Snow

The simple definition of snow in meteorology is 'solid precipitation in the form of ice crystals'. If you've ever seen snow really close up under a microscope, or at least seen pictures of it magnified, then you'll appreciate just how beautiful it is. Nature can be fascinating in so many ways, and how snow forms and transforms our environment is truly stunning. Also, unless we are attempting to travel during a snowy spell, it can bring out the child in many of us.

Snow originates from clouds where the temperature is below freezing and water vapour goes directly from a gas state to a solid state as ice crystals. The number of ice crystals steadily increases as do the frequency of collisions. This process ultimately gives birth to a snowflake. There comes a point when a snowflake is too big and heavy. The updrafts within the cloud fail to keep it suspended, and then gravity wins – the snowflake falls to Earth. If the air from the cloud to the ground is also below freezing, the snowflake will maintain its icy crystalline structure as it falls. If the ground is then cold enough, it will settle without melting. If the ground is a fraction too warm, the snowflake will start to melt. However, if the intensity of snowfall is great enough, the sheer number of snowflakes can counteract any gradual melting right at the surface, so you'll still get snow accumulating. There are a lot of 'ifs' in this explanation, and a reason why is when the air is close to freezing a snowy outcome can become marginal; a challenge for any forecaster!

Is Every Snowflake Special?

Because of the unique way in which snowflakes form, under ever so slight differences in temperature, pressure and humidity, almost every snowflake is different. You are very unlikely to find two that are exactly the same, but there are some common features. All snowflakes are hexagonal and have six sides. Some snowflakes have smooth sides like a hexagonal plate and others have six arms growing from each side, giving them the lattice shape you would most likely recognise as a snowflake. If you took a snapshot of the moment water vapour freezes into an ice crystal at the start of a snowflake's life, they would all look very similar as a hexagonal plate. As this plate then moves around the cloud it will experience infinitely different temperatures and humidity from its neighbours, shaping it further. Indeed, not all snowflakes have the recognisable six points and there are actually eight basic shapes a snowflake could grow into: stellar dendrites, columns and needles, capped columns, fern-like stellar dendrites, diamond dust crystals, triangular crystals, twelve-branched snowflakes, and rimed snowflakes and graupel (see page 57).

The shape a snowflake will take is mostly dependant on the temperature and humidity range. For example, plates tend to grow in drier air than the dendrites, and simple plates and columns grow in warmer conditions (above 20°C) than more complicated branching patterns, which form in colder conditions. As you can see, on a microscopic level, snow is complicated!

The most photographed and symbolic snowflake is the stellar dendrite, which has six arms coming from the central plate hexagon. It is the arms on each snowflake that make them unique and it depends on the path each individual flake takes within the cloud that makes the arms grow differently but also independently from each other, although on a single snowflake each of the arms will look synchronised because they are all growing under the same environmental conditions.

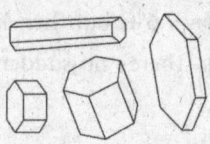
Diamond Dust Crystals

Fernlike Stellar Dendrites

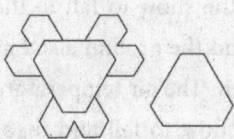
Triangular Crystals

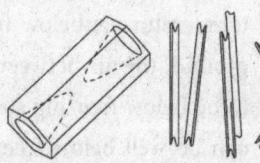
Columns and Needles

Twelve-branched Snowflakes

Stellar Dendrites

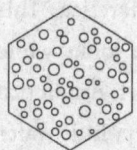
Rimed Snowflakes and Graupel

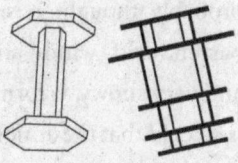
Capped Columns

Types of Snowflake

Can It Be Too Cold to Snow?

This is a question we get asked a lot. Every time the thermometer drops and snow is in the forecast, it seems to come out. The theoretical answer is no, but the practical answer is yes. For example, there won't be a situation where snow is falling while the temperature is at say −6°C but then if it gets colder, the snow suddenly stops. There's no sudden snow on/snow off temperature.

Confused? Let us explain.

Snowflakes are formed in the part of a cloud where the temperature is below freezing. For the snow to fall to the ground, the air between the cloud and the ground also has to be below freezing so it doesn't melt. The air temperature can be well below freezing for the snow to fall and theoretically, how low the temperature gets is irrelevant. What *is* relevant, however, is the humidity. This is the amount of moisture available for the formation of snowflakes. The colder the air gets, the less moisture the air can hold and so there are fewer snowflakes. A better question therefore is 'Can it be too dry to snow?'

One great example of this theory is an area where you probably thought it snows all the time: Antarctica. This barren, cold, windy and icy continent, which looks white and very snowy, is formally classified as a desert. Antarctica is so cold that the atmosphere above it holds very little water vapour so precipitation is very low. When it does snow, it will stick on to the ice sheet and build up over decades and thousands of years.

Is Snow The Most Important Weather on Earth?

While too much snow falling in our towns and cities can cause transport disruption, school closures and general chaos for many of us, it's very important for life on Earth. We don't need (or necessarily) want it in our towns and cities, but we most certainly need it somewhere. Thankfully, there is an abundance (some 46 million km^2) of it around the poles and on the top of mountains every year. Because snow appears white, it acts like a good mirror. This is important because it reflects a lot of the Sun's energy back to space and in meteorology, we call this the albedo. White surfaces have an albedo of 1 and black surfaces 0 and snow can reflect up to 90% of incoming solar radiation. While the amount of snow throughout the year varies, particularly on the land in the Northern Hemisphere, it regulates Earth's temperature by cooling it and keeping the average global temperature controlled. This is one of the reasons why problems arise when global warming melts more of the snow and Earth becomes less reflective.

It is also important on a local scale for communities and ecosystems. People living around mountainous regions rely on the spring and summer melting as their source of available drinking water and irrigation for their crops. For plant and animal life, snowpack can protect the ground during the winter by insulating it and keeping it warm, like a blanket.

ICE

Just in case there was any doubt, ice is the solid state of water. In most cases it forms when water reaches and goes below 0°C. We say 'most cases' because, it has been touched on elsewhere, water can still be a liquid when the temperature is below zero, namely as supercooled water.

Water freezes when the hydrogen and oxygen molecules (H_2O) are so cold that they slow down and hook on to each other to form a solid crystal. One of the most interesting things about ice is that while it is a solid, it is actually less dense than liquid water. We won't go into the details, but on a molecular level, ice has more gaps (air) between the hydrogen and oxygen bondings than it does as a liquid. This is why ice floats on water and doesn't sink. Just imagine if the molecular structure of ice meant that it didn't float, ice would form and then sink to the bottom of the ocean. The current habitats of the ocean floor wouldn't be the same, and there'd be no polar sea ice or icebergs. The planet would look totally different and would perhaps be empty of life forms as we know them.

How Important is Ice?

The collective term for frozen water all over the world is called the cryosphere. This term encompasses all the ice over land and sea, including the Antarctic, Arctic, glaciers, ice caps, icebergs, permafrost and frozen rivers and lakes. Through the year, the total amount of ice on Earth changes as the seasons change. The biggest variations are in the Arctic and Antarctic sea ice, which melts during the summer months. In the Arctic, the winter extent covers around 14 to 16 million km^2 falling to only 7 million km^2 in summer. The Antarctic has a greater range, from around 17 to 20 million km^2 in the winter to only 2 to 4 million km^2 of sea ice in summer. Of course, there are other major sources of ice sitting on top of land, the extensive sheet across parts of Antarctica, where there is approximately 14 million km^2 of ice. Greenland has around 1.7 million km^2 and glaciers of the world cover approximately 726,000 km^2. It is worth pointing out that the Antarctica and Greenland ice sheet hold around 68% of the world's freshwater supply.

You may recall in the snow section earlier in this chapter that the whiteness of the snow, or albedo, is important on Earth as it reflects incoming solar radiation and regulates the global temperature. The albedo is a measure of how reflective a surface is. A black surface absorbs incoming solar energy with very little reflection and would have an albedo of very close to zero. White surfaces will have an albedo closer to 1, which means it is a good reflector. Note how most houses in the Mediterranean are white and the most popular colour of car in the Middle East is white. This is purely because the white reflects most of the solar energy and it keeps the occupants cooler than if they were in a dark/black environment.

Oceans without ice on top have a typical albedo of 0.06, which means only 6% of the incoming solar energy is reflected back into the atmosphere, 94% is absorbed and used to heat the ocean. Sea ice has an albedo of 0.5 to 0.7, which means only 30–50% of the incoming energy is absorbed, keeping the surface cooler. If the ice has a layer of snow on top, the albedo is even higher at 0.9, meaning only 10% of incoming energy is absorbed, with 90% of it being reflected back out to space. These numbers really matter. Greater and extended heating from Spring to Autumn over these icy climes reduces the time necessary for essential snow and ice regrowth in the winter. It's a delicate balance. The overall albedo falls, more heat is absorbed into the system causing accelerated ice melt. The faster the melting occurs, the lower the albedo, a worrying process that is called a positive feedback process. It's hard to know what is positive about this, in the climate system ice melt is its own worst enemy – with ice melt comes more melting.

Earth's surface is made up of 75% of water, of which around 97% is saltwater, with the remaining 3% being freshwater. A significant proportion of this is locked up as ice in the glaciers and ice sheets of the planet. During the summer, the melting of a small proportion of this ice in mountainous regions provides vital supplies to reservoirs as drinking water, as well as agriculture, irrigation, industry and hydropower generation. Because the polar regions are very sensitive to climate patterns and the ice here is on the frontline of manmade climate change, it's an environment closely studied by climate scientists. Ice sheets and glaciers have been around for hundreds of thousands of years, formed by snowfall accumulating and compressing, year after year.

Some glacial ice is several miles thick. In each passing year, the climate would have played a role in how much snow fell and, at a molecular level, the ice chemistry can tell us about the state of the atmosphere. By drilling into the ice sheet and collecting an ice core, scientists are able to look at the thickness of certain layers and bubbles in the ice. This can tell us a lot about the concentrations of greenhouse gases, the length of any glacial periods and the stability of the climate over the last 10,000 years.

By studying ice core, scientists are able to determine how much carbon dioxide, methane and other greenhouse gases there were in the atmosphere for a particular period of time. The presence of ash, dust or other particulates can give an indication of any volcanic activity and how that affected the climate. Looking at the molecular structure of ice within an ice core can inform scientists of the past precipitation, which can in turn be used to reconstruct previous climatic temperatures. Like an archaeologist or palaeontologist digging around in the soil and rocks to gain an insight into a past world, palaeoclimatologists 'dig around' in the ice core to see what the weather was like. The deeper we can extract ice cores, the further back in time we are able to go. One ice core has been drilled in Antarctica which goes back 2.7 million years. It's an astonishing record of past climates showing ice ages and concentrations of carbon dioxide. Many ice cores have been drilled across Antarctica, Greenland, the Arctic and certain glaciers to give scientists as much data as possible to put together a global computer model. Using their understanding of atmospheric physics, climatologists are able to analyse past climates and then use the computer model to help predict future changes in our climate.

How Concerning is Melting Ice?

Earlier, we talked about melting ice changing the albedo of Earth's surface, making it less reflective and absorbing more solar energy. But this is probably not what most people would cite as being the biggest problem with melting ice. Sea-level rise is seen as one of the gravest consequences of global warming and climate change. However, not all melting ice leads to sea-level rise, it depends on where the ice is located.

When Simon was a young lad with aspirations to become a meteorologist, he went to listen to a talk conducted by the British Antarctic Survey about the climate of the large icy continent that extends in every direction from the South Pole. The speakers there demonstrated a simple yet effective way to show the different impacts between melting ice from floating sea ice and ice covering land. An experiment that captivated his imagination and further fuelled his ambitions!

They displayed two glasses; one half filled with water and the other empty. They added a couple of ice cubes to each glass and left them to melt. Many of you may assume, like Simon did, that the water level in both glasses would increase. Once all the ice had melted, the glass that started dry now contained water but the level of water in the other glass that originally was filled half with water hadn't changed. This process can be amplified a million times and it still holds true. The reason why we don't see a discernible increase in the level from sea ice melt is because the mass of ice is already displacing a certain volume of the water. Any melting is just replacing the same volume of melted water that it was initially displacing out. However, there is one caveat to this. There is a difference in density between freshwater and seawater. Freshwater isn't as dense as seawater, so the ice has a greater volume than the equivalent seawater. Therefore, when the freshwater ice melts it adds a greater volume of water than it originally displaced. The difference is very small, but scientists suggest that melting sea ice raises sea level by a mere 0.005 millimetres a year.

The whole point of explaining the physics behind melting ice in freshwater and saltwater is that while the melting of the Arctic sea is obviously detrimental to rising global heat due to lower albedo, sea level rise will be slight. This is not the case for continental ice melt. The biggest and most worrying factor in rising sea levels is the collapse and melting of the Antarctic and Greenland ice sheets. If this were to happen, along with the melting of glaciers around the world, the global sea level could rise as much as 70 metres. The consequences of this happening are immense; hundreds of millions of people would be displaced as about 40% of the world's population live within 100km from the coast.

How Can Something Be Freezing, But Not Ice?

We all learn that H_2O can exist as one of the three states of gas (steam), liquid (water) or solid (ice). It is also commonly known that if you heat water enough, it will turn into steam and if you cool it below 0°C, it'll turn into a solid. This isn't always the case. In meteorology, we can have H_2O still existing as a liquid even when the temperature is down at below -40°C. In this state it is known as supercooled water. We've briefly mentioned supercooled water in other parts of the book, but let's get into some of the nitty-gritty science of it.

At temperatures below 0°C, water will need to attach itself to a nuclei such as a microscopic bit of dust, pollen, ice crystal or pollutant to start off its crystallisation process. When there is a no such nuclei and the water droplet is 'pure', i.e. containing only hydrogen and oxygen molecules, there is nothing to start that crystallisation off. Water won't just freeze all by itself until it gets down to that key -40°C point when crystal homogeneous nucleation occurs. Supercooled water can exist in many clouds, but it is often the mid-level clouds like altocumulus or altostratus that are high enough in the atmosphere for the temperature within them to be below zero, but for them to still exist as non-ice clouds. You can tell the difference as ice clouds are clearly visible as smooth white, wispy clouds really high up in the sky.

Does Hot Water Really Freeze Faster Than Cold Water?

Huh? The obvious answer is 'of course it doesn't'. If you place two cups of water, one hot and one cold, in the freezer then you'd think that under the same rate of cooling, the cold water would reach zero before the hotter water. But there have been experiments where the hotter water actually freezes first and the reason why has quite brilliantly baffled scientists for centuries! You may have also seen that cool experiment when it is really cold outside and someone throws a cup of boiling water into the air, which then freezes instantly into tiny ice crystals. This doesn't work if you throw cold water into the air.

The debate started in 1963 when a Tanzanian student called Erasto Mpemba was making ice cream in class. He made his ice cream by boiling cream and sugar together in a pan but instead of letting it cool before putting it in the freezer like the rest of his friends, he put his mixture straight into the freezer hot. He found that it froze quicker than the cold mix. Baffled by his findings, he asked a physicist, Dr Osborne, to explain what he'd observed. Dr Osborne found similar results in experiments. It should also be noted that Aristotle, the Greek philosopher from the 4th century BC, said, 'The fact that the water has previously been warmed contributes to its freezing quickly; for so it cools sooner.' Osborne and Mpemba published the results together and the effect is known as the Mpemba effect.

Scientists have been split about the effect though as some experiments haven't been able to replicate the original Mpemba findings, but others have. There are many theories as to how the Mpemba effect works, but the problem is that scientists aren't that convinced each of them is the sole answer and so it is still hotly debated.

FOG

In its simplest term, fog is cloud at ground level that causes a reduction in the visibility to below 1,000m. Any reduction in visibility from cloud on the ground is known as mist. This reduction in visibility is because cloud, and therefore fog, is made up of billions of tiny water droplets that get in the way of sunlight passing through and reaching your eye. So, the thicker the fog, the less you can see through it.

There are two main types of fog: radiation and advection. Before we explain the differences, the way fog forms is the same for all types. Air contains water vapour that is invisible to the eye. If you want to prove that to yourself, breathe into your hands a number of times and you'll feel your palms becoming moist with the water vapour changing state from gas to liquid. As air cools, it will reach a point where it can no longer hold the moisture and will condense into tiny water droplets. The temperature in which it does this is called the 'dew point' temperature and is very important for weather forecasters to know. Just like normal air temperature, the dew point can vary depending on the time of year and where you are. If the temperature of air falls below around 2 degrees of its dew point temperature, that becomes the critical point for when the vapour turns into water. Fog is then essentially when there are enough water droplets suspended in the air near to the ground to reduce the visibility.

There are differences in *how* the air is cooled enough to its dew point temperature. The most common type of fog you can get in the winter time is radiation fog. One of the most important weather conditions we need to consider is the wind. For radiation fog formation, you need a very light wind. Too much wind and the bottom part of the atmosphere, called the boundary layer, gets too turbulent and mixed (though you do need a little bit of this). It is called radiation fog because the air close to the ground cools by thermal radiation. The best way for the air to cool this much is when there are clear skies and enough heat is allowed to escape into the atmosphere (too much cloud and it acts like a warm blanket at night, keeping temperatures higher). As the air continues to cool, it may fall within 2 degrees of the dew point temperature and the fog will form. The fog will eventually clear away during the morning as the Sun starts to heat the top of the fog or ground. It can also clear if a stronger wind starts to develop.

Advection fog is a type of fog that occurs most typically over a body of water, such as a lake or sea. The physical process of air at the surface cooling and condensing is the same. However, in the case of advection, you have warm, moist air moving in over the top of relatively cooler air above the water's surface. This means the warm air will start to cool to the point where it can't hold the moisture and creates water droplets. Fog is formed. Unlike radiation fog, where you need light winds, advection fog can occur in moderate to strong winds as it is air moving across the water's surface. The wind can often move the fog inland so in some circumstances coastal areas can be foggy whereas a few miles inland is completely clear. Advection fog can last for many hours and days if there is a constant feed of this warm, moist air coming in over the top of the cooler waters.

Upslope hill fog occurs in a similar fashion to advection fog, where warm, moist air is forced up a hill or mountain. As that air rises, it will cool and condense once it reaches its dew point temperature. This is also essentially cloud forming on the side of the hill or mountain but as someone standing there, you'll just notice a reduction in visibility from fog.

Sometimes you might hear in a weather forecast the presenters or meteorologists talk about freezing fog. This is very simply fog that occurs when the air temperature is below freezing. The water droplets comprising the fog may not freeze as they can exist as supercooled water. The water will only start to freeze once it hits a surface, which will set off the freezing process. So, while this type of fog will give the usual reduction in visibility, you may also see white deposits of ice forming – this is known as rime.

Fog is actually quite a challenge to forecast accurately in any particular location because it can be very localised. In some weather situations you can get a good idea that fog formation may be widespread over a large area or on other occasions, it can be very patchy and dense. This type of fog is one of the most hazardous as you could be driving in clear conditions but then suddenly hit patches of thick fog.

The first thing a forecaster would consider is the broader weather set-up. For radiation fog to form we've established you need light winds. Secondly, we need to know what the dew point of the air mass is across an area and then forecast how low temperatures are expected to fall. If it is within 2 degrees of the dew point, then there is a good chance of fog formation, especially if there is enough moisture in the boundary layer.

Valleys are more prone to radiation fog and they might be areas where we can be more confident in forecasting fog. This is because the temperatures here can be lower than surrounding areas and therefore more likely to drop to the air's dew point temperature. Forecasting when fog is expected to clear is equally tricky and as an operational forecaster in the Royal Air Force, Simon had many frustrating mornings briefing aircrew about when the fog might clear so they could get out and fly. One of the ways in which fog clears is for the air temperature to rise, which will then 'burn' off the fog by heating the water droplets until they evaporate back to water vapour. Forecasting techniques would be used to establish the temperature at which the fog would clear and then to figure out what time that air temperature was expected to be reached. Without using professional forecasting techniques, as a very general rule of thumb you can take the month you are in and 'add two' for the time of fog clearance. For example, if you are in September, that is the ninth month, add two to make eleven therefore eleven o'clock fog clearance. Note that this doesn't always work!

Another way fog can clear is if the wind strengthens, which will mix up the fog layer and disperse it. Cloud at night can also clear fog if it rolls in above the fog layer as it will warm the ground slightly. If cloud rolls in mid-morning over the fog, it can prevent it from clearing as the Sun can't warm the top of the fog layer or ground enough to burn it off. Fog is definitely a tricky customer.

For forecasting advection fog, including sea and hill fog, we will again look at the bigger picture in terms of weather set-up. One question might be, is there a warm front bringing warm air in across the UK? If there is, we'd again assess the dew point and forecast temperatures. In a lot of cases as a warm front passes in from the west or southwest across the UK, there is normally a lot of low cloud and moisture available for advection fog to form around coasts and upslopes. This type of fog formation is often a bit easier to predict than the radiation fog.

What is a Peasouper?

On days when there is really thick fog you might have used the term 'peasouper'. It originates all the way back to 1820, when an artist at the time described the London air as being as thick as pea soup. In those days it was often not just fog that caused a reduction in visibility, because of the burning of coal in houses for heating, big cities such as London had major pollution problems and the air would be filled with thick smog. On days when there were light winds with fog developing, this weather mixing with the smog would make visibility extremely low. The air would be so polluted that it was poisonous with sulphur dioxide and soot, and it led to the deaths of vulnerable people such as the elderly, young and those with respiratory problems. Peasoupers became so bad that after London suffered its worst smog event, with around 12,000 deaths being attributed to the conditions, the Clean Air Act was introduced in 1956. While today the UK doesn't have the same smog and pollution issues, any time fog is particularly thick, we might call it a 'peasouper'.

Foggy facts

- Fog is when the visibility goes below 1,000m. Mist is when the visibility is more than 1,000m and less than 8km. Haze is when the visibility is between 2 and 10km, but the humidity above is less than 70% (i.e. the air is too dry for the reduction in visibility to be due to water droplets).
- The foggiest place in the world, where there is a fog for over 200 days a year, is in Grand Banks, near the Canadian island of Newfoundland.
- In 2006, across the Southeast of England thick fog and freezing fog persisted for days in the lead up to the Christmas getaway, causing chaos. Hundreds of flights were cancelled over a five-day period.
- Locals in the Atacama Desert in Chile, one of the driest places on Earth, use nets to 'catch fog'. As fog rolls in from the coast, water droplets become stuck on the nets, where they then drop into a gutter and get collected as water. This can provide as much as 70 litres of water a day.

- Thick fog helped George Washington against the British when on 29th August 1776, weeks after the signing of the Declaration of Independence, Washington and his army were trapped in New York City with the British closing in. Thick fog over the British position meant they couldn't advance. While they were sitting around waiting for the fog to clear, Washington was able to sneak out of Brooklyn with his 9,000 men without a bullet being fired.
- Fogbows are rainbows that form in fog. When the Sun is behind you, the light reflects off the water droplets in the fog in the same way it does in a rain shower. You may often only see very faint colours because the water droplets are smaller in fog.

LIGHTNING

How Hot is Lightning?

While you may not see lightning very often, it might surprise you to know that on average, around the world, there are approximately 100 lightning strikes hitting Earth's surface every second. The vast majority of these lightning strikes occur around the tropics in areas known as 'hot chimneys', where there is enough heat energy to produce big cumulonimbus clouds and thunderstorms on a regular basis.

Our atmosphere contains an electrical charge and even in fair weather, you can measure that charge at the surface. Once cumulonimbus clouds develop, the electrical charge grows and we get lightning strikes. As we know, the cumulonimbus cloud also has severe up- and downdrafts. The mixture of water droplets and ice in the cloud experiences extreme turbulence from these drafts, with numerous collisions creating friction. This friction creates the static charges within the cloud. There will be a mixture of positive and negative charges that will naturally separate. The positive-charged particles will gather towards the top of the cloud while the negative charges head to the bottom, just like a battery with a plus and a minus. Once there is enough charge, the atmosphere will want to equalise itself, so you get a very intense and quick lightning strike going from the negative to positive charge. Most of the time this strike will be upwards to the top of the cloud, which is called intra-cloud lightning. The ground is also made up of positive charge so if the electrical potential is strong enough, the lightning strike will head down to Earth and is called cloud–ground lightning.

Technically, lightning itself doesn't have a temperature as it is the movement of electrical charges. But as the lightning passes through air or another material, it will heat this material up. Air will get very hot as lightning passes through it, forming a channel in the air, with the temperature rising to almost 27,500°C in a split second. That's nearly five times hotter than the surface of the Sun (5,500°C). It is this process that generates the thunder sound during a storm. Creating the channel, coupled with the heating, suddenly compresses the surrounding air and leads to massive shockwaves. It is these extreme vibrations in the surrounding air that travel to our ear and give us the sound. If the lightning strike is nearby, it might sound like a loud whip and crash. If the lightning is much further away, then as the sound travels, it gets distorted in the air and when it reaches us, it will be more like a longer grumble.

Could We Harness the Power of Lightning?

The numbers are impressive: a single bolt of lightning contains around a billion joules of energy, which is enough to power a home for a month. So, theoretically, it would be a brilliant idea to collect that power and use it as a renewable energy source. However, there is a very good reason why this doesn't work. For starters, while the amount of energy available in one strike is huge, it is delivered in a split second. We don't have the engineering techniques to be able to handle this surge of energy in a short space of time, store it and then release it as an energy source over a longer period. The other major problem is that lightning is extremely sporadic and it is almost impossible to specifically pin down where it might hit. Most of the lightning strikes occur in areas of the tropics, where population density is sparse. Even if it was possible to harness the power from all the lightning striking Earth, scientists have calculated that every year we would only be able to power around 8% of USA households.

Where Does Lightning Strike More Than Once?

Lightning occurs most frequently on land around the tropics as this is where there is more heat energy and therefore convection to produce the towering cumulonimbus cloud necessary to generate thunderstorms. The most lightning strikes across the world happen over Central Africa, Central America and the Asia Pacific. Within these regions, it is over mountains where they are most frequent. Mountainous terrain enhances the upward motion of air, so the convection is boosted further by local topography. If you add a lake into the mix as well, the additional available moisture can easily be transferred above, populating the sky with thunderstorms and dramatic lightning displays. In Venezuela, where the Catatumbo River meets Lake Maracaibo, there are on average 250 lightning strikes per square kilometre every year, with around 28 lighting flashes each minute. Another electrically charged place is the mountain village of Kifuka in the Democratic Republic of the Congo, where there are 158 lightning flashes per square kilometre each year.

CLOUDS

WHERE DO CLOUDS COME FROM?

Every day, clouds race across the sky, gracefully sweeping in, only to disappear; their relationship with the Sun, land and sea determining their ever-changing form and life-span. But not every part of Earth's atmosphere is a breeding ground for clouds. As a whole, there are seven distinct layers of the atmosphere, each with unique properties to protect the planet. The lowest slice transports water, in all its guises over land and sea, sustaining the richness and diversity of life.

The troposphere, the lowest layer of the atmosphere, is home to life-affirming oxygen, abundant nitrogen and a small yet vital mix of carbon dioxide, water vapour and other greenhouse gases. It's here that the weather machine continually flows, distributing heat and water across the globe. The most essential dynamic in this zone is that air cools with height. This dynamic alone allows convection, advection and condensation to happen – the fundamental tools of cloud formation. This happens at every level of the troposphere, with each part interacting and mixing, sweeping upwards and dipping to the oceans.

The Recipe for Clouds

Ask a class of five-year-olds what clouds are made of and at least one will shout out 'fluff!' And they certainly do look like that! But, in fact, the white stuff that floats in the air is a product of millions upon millions of tiny cloud droplets that eventually battle for space as more ping into existence. Ultimately, they become one assuming the guise of a white fluffy cloud.

Putting fluff to one side, the basic ingredients for a cloud are water vapour and heat energy. Heat transports air loaded with water vapour into a cooler environment through advection or convection, and then condensation transforms water vapour into tiny water or ice droplets. Condensation nuclei, such as salt and dust, are also a key component of cloud droplet formation. Tiny condensed water molecules gather by latching on to relatively larger aerosols. Condensation nuclei are about 1 micron, whereas these tiny water molecules are in the order of 0.0001 microns. As more water molecules attach, it doesn't take long for cloud droplets to form, each hosting its own aerosol. The lightness of cloud droplets mean they remain suspended and become the body of a cloud as the feed of water vapour continues.

The Cloud-making Process
Convection: the rising of air, due to heat;
Advection: the horizontal transport of heat by the flow of air;

> **Condensation:** when water vapour (gas) cools and changes into water (liquid);
> **Rising and falling:** the sun heats Earth at different intensities. This flow of air that rises then falls, recreating large zones where air pressure is high (descending air) and low (rising air). These currents of air carry water and heat. It is a perfect energy system that moderates extremes in cold and hot, dry and saturated. Rainforests, deserts, tundra, icy poles and verdant mid-latitudes exist all due to Earth's weather system, keeping the planet rich in diversity with blues, greens, whites and browns, and every colour in between.

THE CLOUD PARADE

A classic set-up that plays out across the sky is the weather frontal system with a warm front (depicted on synoptic charts by red semicircles on a red line) followed by a cold front (blue triangles on a blue line). Each part of this process has its own clouds and is a good indicator to the observer of which part of the front is currently overhead. With differences in air density, cold (denser) vs warm (lighter), air masses don't just merge, the warmer air rises over the top of the colder air, where they initially have distinct boundaries before starting to mix. This can be seen in the progressive cloud structure. It all starts with the layered stratiform cloud before clearing into the heaped, unstable, cumuli cloud.

- *Cirrus*: the high veil that allows the Sun to shine through. Initially, the warm air skirts alongside the cold air, almost like air blowing over a block of ice that it bumps up against. The first mixing happens at the top of the troposphere, evident from the veils of cirrus that stretch across the sky. These thin, wispy clouds made of ice crystals can sometimes be a good signal that the clouds will thicken further with the eventual onset of rain.
- *Altostratus*: medium-level cloud, covering the Sun. Slowly, the warm air penetrates the cold air mass and progressive layers of cloud extend across the sky. Altostratus clouds aren't weather-making but are a precursor to the lower cloud that eventually brings rain.
- *Nimbostratus*: 'stratus' meaning layer and 'nimbo' meaning rain-bearing. At this stage, a thick layer of cloud and moisture dominates from low to high up in the atmosphere. These are the rain-bearing clouds that hang heavy and low in the sky. They follow the higher clouds and indicate the onset of rain and poor visibility as the trapped warmer air eventually mixes with the lowest layers of the atmosphere. We call this moist, mild zone the warm sector.
- *Cumulus*: as the nimbostratus clears, the skies will brighten, indicating a clear transition from the warm air mass to a return of colder and clearer air. At this stage, the air is now unstable and allows cumulus clouds to start building in the sky. They signpost a passage of the cold front or a return to colder air. These have a far

more defined shape and are discrete bundles of cloud allowing the typical 'sunny spells' to come alive. As cumulus clouds gets bigger, into 'humilis', 'mediocris' and 'congestus', they can start producing rain showers. These are also coupled with a veering, squally wind – a testament to the transition from warm to cold. Cumulus clouds don't necessarily form from lower levels, they can also develop higher up in the troposphere and indicates another unstable portion of the sky. These are known as altocumulus in the middle part of the troposphere and cirrocumulus at the higher levels.

Satellite imagery captures beautifully this family of clouds as they develop and track. They show as swirls that curl to the centre of the low pressure and splay out hundreds of miles. Easily recognisable in such images is the smooth form of the layer cloud and the open, lumpy pattern of the cumulus clouds, a clear indicator also of the change of air mass.

In a nutshell, these large weather systems are a result of a battle between different air masses – imagine a huge mass of cold air surging south from the icy northern reaches. Another mass of warm air rich in tropical moisture pushes north. At some point over the ocean they will clash; two vast volumes of air, each with their own characteristics. It's north vs south, cold vs warm and poles vs tropics.

The layered cloud family may not strike wonder in the hearts of those who photograph, paint or just stare at the sky, but their partnership with the cumuli in some ways completes the story of cloud development. When they march together, they tell of a journey so immense and so powerful. A marrying of two extreme sources – the tropics and the poles. Low pressure and its associated frontal systems are vital in redistributing heat and water across the globe. Each cloud plays its part, from the thinnest veil of cirrus – the first sign that the air is beginning to mix and of what's to come – to the boldest cumulus bringing up the rear and spelling the transition from rain to showers and clearer, colder air.

AT WHAT POINT DO CLOUDS RAIN?

Perhaps an even more tricky concept to fathom than what clouds are made of is how does the rain remain suspended in a cloud? A lot of question marks float around these beautiful suspended water carriers in the sky.

What we do know is that collisions between cloud droplets will produce larger cloud droplets. This process is called coalescence. As further collisions happen, cloud droplets grow into rain droplets. It is then the strength of local currents within the cloud, that keep it suspended despite gravity, will determine if and when the rain falls. At a certain mass, the rain will be too heavy for the forces within the cloud to hold it, and gravity wins.

WHY DO CLOUD BASES ALL SEEM TO BE AT THE SAME HEIGHT?

As air rises, it cools and with cooling, there is a threshold where water vapour can't hold the existing volume of water vapour as a gas any more and at this temperature the air begins to condensate and clouds start to form. This is known as the Lifting Condensation Level or LCL, the base of the cloud. It's when the air has become cold enough for its moisture to condense into tiny liquid drops of water that eventually form clouds. This temperature is known as the dew point and varies from area to area, but locally, the air tends to exhibit similar properties, resulting in a similar LCL, and so clouds within an area tend to form from the same base level.

HOW LONG DO CLOUDS LAST FOR?

Clouds last as long as there are enough internal currents to keep convection, condensation and coalescence going and while there is a feed of heat and water vapour into the clouds. This continual process is one of the reasons why all water droplets don't fall all at the same time. Another consideration is the mixing of drier air and evaporation of water deposits from the cloud. Clouds can be warmed by solar radiation from the sun and longwave radiation from the ground. Fog or stratus cloud on cool mornings can often disappear by the time the sun rises and daytime heating seemingly 'burns' the cloud away by evaporation. When this feed weakens, the clouds shrink and disappear. For some clouds, this is when the daily temperature falls during sunset (loss of convection). For other clouds, it's when two distinct air masses have mixed and pretty much neutralised their energies (loss of advection).

It's All About Convection

While it may not be visible to the naked eye, the air comes alive during a hot sunny day. The Sun's light energy is absorbed by the ground. The energy is re-emitted into the lowest layers of the atmosphere as infrared energy or heat. When this air is warmer than its environment, thermals – a glider's best friend – gracefully rise through the air. As the Sun continues to beat down, the lower layers of air slowly warm up further. This process is called convection and it describes the heat transfer of a fluid, such as air, causing it to rise when it's warmer and therefore becoming less dense than its environment. Local rising heat may be important but the key fuel that any cloud needs to come into life is moisture. The passage of air over water will allow for the water vapour content to increase and interestingly, the higher the temperature, the more volume of water vapour a body of air can hold. With the right environment of hot air rising and cooler air at height, the stage is set for an explosive event.

As the drama unfolds in the skies above, meteorologists, weather enthusiasts and storm chasers talk CAPE – a little buzzword with huge expression. CAPE refers to the Convective Available Potential Energy of the storm; the measure of how much energy is available to fuel the storm. It indicates how viscous the storm will be and its ability, ultimately, to spawn into a supercell and generate tornadoes.

How Can You Tell a Storm is Coming?

The weather forecast might predict a risk of thunderstorms during the day, but on many occasions you may not even see one. While forecasters can predict that the correct atmospheric conditions are there for cumulonimbus clouds to form and therefore a thunderstorm to develop, it's much harder to say where exactly they will form. It's almost like watching corn popping in a saucepan: the conditions are perfect for the kernels to start popping but you can't predict exactly which spot and which corn will pop first. On a day that might be primed for thunderstorms, there are certain signs you can look out for to tell if a storm is going to affect you.

The first thing you will notice is the development of the cumulus cloud, the typical summer's day fluffy cloud in the sky. With enough heat and moisture in the atmosphere, the cumulus clouds will develop and grow into cumulus congestus. The clouds at this point will be getting darker and darker as they get thicker. The next stage is when these clouds grow further into cumulonimbus. If you start to hear thunder, then you will know that the thunderstorm is close by. As the thunderstorm gets closer you may also start to see the lightning before the thunder – even more so at night when the lighting flash is more noticeable. This is because light travels faster than sound so if the storm is still some distance away, you will see the lightning before the sound of the thunder reaches you.

As a child you may remember doing the little trick of counting between the lightning flash and the thunder to find out how far away the storm is. This is actually a pretty good way of establishing your distance from the storm. For the proper mathematical way of doing it, count the number of seconds between lightning and thunder and then divide that time by five to give you the miles away. Obviously, if the flash of lightning is followed straightaway by the thunder, the centre of the storm is right over you.

While the above describes what might happen on a day when thunderstorms develop locally, storms can form in other, more dramatic, ways. During a 'Spanish Plume' event, warm, moist air flows off the Spanish plateau and can travel northwards through France and towards the UK. If this happens at the same time as cooler air from the Atlantic is pushing in from the west across the UK, it creates a huge amount of instability within the atmosphere. Air at the surface can easily rise high into the atmosphere, morphing into a cumulonimbus cloud. A precursor to this type of event is spotting a cloud in the sky called altocumulus castellanus, lovingly referred to by meteorologists as Ac Cast. We say 'lovingly', because it's quite a beautiful cloud that curls and spirals high in the sky – not a common sight. A stunning indication that instability is playing out in the mid-atmosphere, and if the lower part of the atmosphere also becomes unstable, as temperatures at the surface start to rise, it can result in the whole atmosphere destabilising very quickly, with towering cumulonimbus clouds forming.

Some people claim that they can sense when a storm is coming. This is a weird one as Simon would argue that in his days of forecasting at military bases, he would spend quite a bit of time outside looking at the sky and observing the weather. He certainly thinks that he could 'feel' when a storm was about to hit (he did have the benefit of other data to base that on). But the science behind such a feeling is certainly tricky and while some researchers believe there are some things going on in our bodies, there are still unanswered questions. Some of the theories are that as pressure drops with the onset of a storm, the fluid around our bones can respond, particularly for those people who have arthritis. If your nose is sensitive enough, you may also be able to smell a storm coming. Having read the section on what rain smells like, you'll know that you could smell Petrichor when it rains. Sometimes this smell can be transported many miles ahead of the storm on gusty winds. With an increase in electrical activity within a storm, more ozone is created by lightning. Ozone comes from the Latin 'to smell' and has a distinctive smell that you might recognise in a photocopier. With more of this ozone being generated and being wafted in the winds ahead of the storm, this may be the first thing you sense before a storm arrives.

CUMULUS CLOUDS

Until recent technology mapped and projected atmospheric processes, watching the skies was an essential daily routine. Spotting changing shades, shapes and shadows gave away important clues of what would happen next. The glorious swirling mass of certain clouds populating the sky during particular weather days illuminates the invisible processes that ultimately manifest in weather.

Shape and Height

If air currents are the first responders to the change in heat and water, then clouds solidify and magnify that change. Wherever air moves and interacts with different environments so clouds will form, and are the most visible and tell-tale signs of what will happen next. The art of observing clouds begins with recognising their shape and height, and these two factors alone speak volumes for when it will rain and when the Sun will reappear.

If the cloud has a good bubbly shape, then it's because the local air is warming quickly and rising into cooler air. This means that the cloud is locally produced and we refer to the rain that these clouds produce as showers. This type of cloud is the family of Cumulus clouds, and although it's the lower clouds that form showers, cumulus clouds can form at any level of the atmosphere.

Cumulus clouds or shower clouds tend to form over land when the Sun has enough strength to heat the land, so usually from early spring (or even late winter). During the coldest months, the Sun is weaker and the land remains cold, so these clouds form when over relatively warmer seas, especially when the air aloft is cooler – a cold northerly wind would provide such a scenario. During the winter months, the coast will be affected by more showers, but there isn't the energy inland for these to be homegrown – although with a strong wind, the showers can penetrate the colder inland areas. In winter, more often than not these showers will be a mix of rain and snow, or simply snow.

Colour and Shade

If the clouds are bright white and you can almost see their tops, they will be fair weather features, occasionally blocking out the Sun but doing no more than dancing across the sky before drying up and disappearing or spreading out into a more shapeless form. The darker the cloud and the more they loom up into the heavens, the greater the chance of heavy weather. The processes of rising currents, condensing into water and ice droplets, will result in stronger gusty winds and heavy downpours. If these clouds keep on growing, with dark bases stretching miles up and spreading out, then expect lightning, thunder, torrential rain, hail and squally winds. If they merge into one seething mass of darkness and start to rotate, it's time to run for the hills, as tornadoes may be next. The shape of the clouds is a good first indicator of what's going on in the atmosphere.

The Cumulus Family
Cumuli means 'heap'.

Cirrocumulus: tiny heap clouds that form at about 25,000ft;
Altocumulus: form at about 10,000ft;
Cumulus: form below 7,000ft and produce rain showers – these are mostly short-lived;
Cumulonimbus: the biggest of the family, extending from the base to the top of the troposphere, these produce thunderstorms.

Cumulo – Heaped; Nimbus – Rain-Bearing

These majestic clouds loom above every other cloud at eight to 13km (five to eight miles), but unlike their flatter cousin, the stratocumulus, most don't last long. The cumulonimbus or thunderhead makes an explosive entrance and although not always, generally runs out of steam and disappears within the hour.

The explosive nature of a cumulonimbus cloud can be seen in its anvil shape; this heated air loaded with moisture surges to the highest part of the troposphere, where it hits an invisible ceiling. It bumps into the stratosphere, the next layer of the atmosphere, where the air is warmer than below and prevents development of cloud. So, instead of progressing vertically, these currents fan out in all directions, spreading the cloud away from the action, and give the cloud its notorious anvil shape.

There are hot regions that rarely see a cloud in the sky – namely deserts, savannah and, even in places where the landscape is rich with life, hot days without clouds can be common. Thermals still develop but without high humidity air clouds are unable to form. Rising air through the day doesn't necessarily create clouds, there needs to be a feed of moisture. However, during the afternoon close to the coast, a circulation can develop, with cooler, moister air from the sea replacing the rising air over the land, causing a local low pressure. This intensifies further if the onshore sea breeze clashes or converges with the prevailing wind from a different direction. A convergence line of wind forces air upwards with a huge might and now all the ingredients are there, and after a dry and sunny day, clouds appear, they build and then the storm breaks out.

What Goes On Inside a Cumulonimbus Cloud?

Cumulonimbus clouds are the biggest and most violent types of cloud. Not only are they the precursor to thunder and lightning, but they can also evolve into supercell storms and spawn tornadoes, or evolve so much that they produce what we call 'daughter cells' and eventually grow into mesoscale convective systems (MCS). These systems are greater than 100km wide and produce extreme winds including tornadoes, intense rainfall, hail and frequent lightning.

We can tell a lot about what is happening inside a cumulonimbus cloud by looking at Doppler radar. As well as analysing the precipitation coming out of the cloud, Doppler radar can measure the speed and direction of the precipitation within the cloud. This tells us how violent the up- and downdrafts are in the cumulonimbus cloud and can even show forecasters the tale-tell signs of tornadic development.

The interior of a cumulonimbus cloud is like no place you'd ever want to visit. Violent vertical winds would throw you around like a washing machine and big hail and inter-cloud lightning strikes would also be life threatening. If you were able to survive all of that then the sub-zero temperatures, at the very least, would give you a bad dose of frostbite. This is why airplane pilots avoid what they call 'Charlie Bravos' (CBs) at all costs.

Sometimes fate works against even the most competent of pilots. There is a remarkable story about a pilot, Lt Col William Rankin, who had to eject from his fighter jet when he suffered engine failure at 47,000ft. Thankfully, he was wearing an oxygen mask otherwise at that altitude he would have suffocated. As he fell through the sky, he went straight into the top of a cumulonimbus cloud, where the temperature was around -50°C. He was hit by severe winds, lightning, hail, rain and thick black cloud all around him. His parachute was designed to open automatically at 10,000ft, but after around five minutes, when he thought he should have reached that level, it hadn't opened: the updrafts were counteracting gravity by throwing him back up the cloud at times. Eventually, the parachute did open, but that made matters far worse for Rankin as the updrafts caught the parachute and he was being pulled back up the cloud for the umpteenth time. All the while, as he was going up and down within the cloud, lightning was still going off all around him, accompanied by a deafening noise of thunder. He was getting frostbite and the water vapour was so thick, he felt like he was drowning. After about 40 minutes of defying death in the cumulonimbus cloud, he finally came out of the bottom of the cloud and fell with a more or less intact parachute to the ground. He had to spend a few weeks in hospital recovering from frostbite, decompression sickness and a number of bruises over his body.

MEET THE STRATIFORMS

The Stratiform Cloud Family
Stratus is Latin for 'layer'.

Stratus: low and depressing (drizzle);
Stratocumulus: a layer cloud that exhibits a more wave-like form low in the sky (doesn't normally produce rain);
Altostratus: blocks out the Sun, middle level of troposphere (7,000 to 10,000ft), opaque;
Cirrostratus: veil of high clouds, so thin they do not block the sun or moon. They are made of ice crystals and exist at six to 12km (20,000 to 40,000ft).

Layers of cloud exist at every level of the troposphere. High up, they are made of ice particles. At the lowest part of the sky, they are almost formless and produce what many would describe as 'nuisance

> drizzle'. That's a cliché comment that makes weather forecasters of old wince.

Enter Advection and the Stratiform Family

A cloud that covers an expanse of sky and has less shape and appears flatter than others probably belongs to the Stratiform family. Stratiform clouds tend to form due to advection or the flattening out of cumulus clouds that are 'capped' by an inversion (when temperature suddenly rises with height). While cumulus clouds are locally produced via convection, stratiform clouds tend to form due to advection. These clouds are sometimes thick, but there is little upward motion as they instead extend horizontally across miles of sky. When these clouds form on a grander scale, they produce rain. This type of cloud sticks around for longer, because it covers a greater area and are less influenced by local conditions happening at ground level.

CIRRUS: THE WEATHER IS COMING IN

Look to the sky, watch for the veil, see it thicken and note how quickly because this can be a clever indicator that rain is on its way. Earth's spin will add momentum and energy into the system, winds may whip up and lash out as the cloud base lowers and the rain arrives. The veil of cirrus clouds tells the discerning observer that rain and wind are 500 miles away, six hours, maybe less, maybe more depending on how quickly the skies unfold.

WILL IT BE FROSTY TONIGHT?

After a sunset, the land loses its source of heat from the sun. Under a vast open sky, the day's warmth can readily escape, mix with the cooler air aloft and cool the surface down. Thick cloud is a good insulator of heat during the night – it acts as a blanket, trapping the day's warmth in the lowest layers of the sky. This may be the difference between a frost or a frost-free night. During the hottest days of summer, cloudy skies can hinder the most oppressive heat escaping and can keep the humidity higher and the air at an uncomfortable temperature. Starry skies will lead to frosty mornings, dewy dawns and sometimes foggy starts. Under normal conditions, when there's not an invasion of a different air mass, the lowest temperatures are half an hour after sunrise, as the land continues to lose heat even after the Sun begins its ascent into the sky. At this point, there is a lag time while the sunlight works its way into the ground and performs that important energy transformation to heat, allowing the air to slowly warm again.

HOW DID WE USE TO PREDICT CLOUD AND RAIN?

Four Great Weather Lores
Red sky at night, shepherd's delight. Red sky in the morning, shepherd's warning
This delightful rhyme gives great foresight, *sometimes*. The mechanism is light scattering, or formally called Rayleigh

scattering, by dust particles in the atmosphere that scatter the blue light and leave the longer wavelengths of red light. Red sky at night occurs when the Sun sets and illuminates the sky from below. It suggests quiet and warm weather and a good chance that the morning will be fair. As most of the UK's weather comes from the west, the Sun being visible as it sets reveals that there is no cloud coming from there. 'Red sky in the morning, shepherd's warning' suggests the fine weather, or high pressure, is now to the east and therefore there is a higher chance of inclement conditions arriving soon.

Ring around the Moon? Rain real soon

Clouds give away a lot about imminent weather. A ring around the Moon is a stunning image of the reflected light penetrating the delicate upper layer of cloud, cirrus, where ice crystals allow light to shine through. There are some clever optical physics at work to create a halo around the Moon. The process is called refraction, in which light is broken down to its constituent colours by the ice crystals. All other clouds would shroud the Moon. But, cirrus clouds are indicative of approaching weather hundreds of miles away. If you see a sheet of cirrus, it may thicken and the cloud base deepen to eventually bring rain.

Clear Moon, frost soon

Except for the summer months, when many places across the mid-latitudes are frost-free, this saying is a good indicator of a frosty night. On a typical day, the Sun's light brings heat to the lower atmosphere. Once this source of heat is lost at night-time, the air cools. However, a layer of cloud acts as a good insulator to trap the day's heat in the lower atmosphere. Without cloud, warm air radiates away and it cools down. During winter and the months closest to winter, this cooling results in air temperatures falling close to or below zero, so clear skies are a good sign that it will be a frosty night.

Pine cones open up when good weather is on the way
There is some sound science behind this. The opening and closing of the pine cones is related to how humid the air is. Each leaf of the cone is made up of scales, in dry weather these scales shrivel up and the cone becomes stiff as it opens up. Thereby exposing its seeds to the air where the wind can pick them up and spread them out. High-humidity air has the opposite effect on the pine cone where the scales absorb moisture, gain elasticity and are able to close up to protect their seeds in their natural position. Therefore, when pine cones open; dry weather on the way, when closed; wet weather is on its way.

THE BIGGER ATMOSPHERIC PICTURE

As meteorologists, we tend to compartmentalise regionally weather and climate conditions. From deserts and rainforests to the mid-latitudes, and even smaller zones like islands and highlands. The movement of air does not recognise borders or boundaries. Even the greater leaps across the highest peaks are not total barriers to this continual flow.

Large-scale circulations of air sweep, curl, duck and fly in such a diverse way that all corners of the Earth are completely connected. This connection between dry and moist, cold and warm goes deeper; the air interacts with the oceans, it skims the surface, continually morphing between liquid and gas. Violent winds churn waters beneath, casting their atmospheric turbulence into the depths of the darkest abyss while the more steady long-distance winds alter the course of ocean circulation patterns not only horizontally but also vertically, and these in turn modify local weather patterns.

Looking down from the sky, the oceans exude a glassy sheen that seems hardly penetrable, yet beneath, the constant flow of water holds incredible intent; traversing the length and breadth of the planet, where even the coldest Antarctic Bottom Water occasionally rises upwards, splaying its most frigid fluid into the upper echelons of the sea and occasionally, the air.

The atmosphere and oceans are not mutually exclusive; they co-exist and re-enforce. Air reacts swifter to climatic change, while the seas lag behind, yet through millennia their stories intertwine. Ice ages, El Niño (see also pages 113–17) and monsoons all owe their genesis and development to both. The story of the atmosphere can only be told by casting the oceans as its co-lead. It is here among the air and sea, where the grand influences the local, and where the truest artforms of the ocean and atmosphere are realised. This is what we call the Bigger Atmospheric Picture.

HOW DOES AIR FLOW AROUND THE WORLD?

Before this can be answered, it's important to bring together some fundamentals of air flow. Firstly, air will naturally flow from where there's more air (high pressure) to where there's less air (low pressure). Secondly, due to the spin of Earth, air tends to be deflected as it moves from highs to lows. This is known as the Coriolis force, an apparent force due to Earth's rotation, the air doesn't flow in straight lines from high to low as it's deflected. It circulates down and out of a high pressure area and then circulates in the other direction and towards low pressure. The direction depends on the hemisphere. In the Northern Hemisphere, air circulates anticlockwise around a low and clockwise around a high. The converse is true in the Southern Hemisphere.

Warm, moist air will rise, and keep on rising if its environment at height is cooler. This works well when there is a clash of cold and warm air masses. The properties of these two air masses are different; warm air is less dense so as these two air masses twirl around each other, the warm air is forced upwards en masse. This slight disturbance in the weather pattern becomes a defined circulation as the Coriolis force continues to exert its influence.

Eventually, this rising air cools and condenses, producing cloud and rain. At the surface this rising air is replaced, streaming in from all directions, rising as it circles the centre of the low pressure. This replacement of air continues until the rising air stops, which happens when the environment becomes homogenous (i.e. there is no temperature gradient with height). However, back at ground level, the rising air reduces the surface air pressure, resulting in zones of low pressure.

Zones of high pressure contain more air and this air is sinking. Air that descends does not tend to produce cloud, although it can trap cloud already there. The air flow circulates outwards and clockwise in the Northern Hemisphere. High pressure feeds low pressure and air naturally flows from where there is a lot to where there is less – it's fluid, and acts like any other fluid.

What is the Weather Associated with Lows and Highs?

Low pressure comes in all strengths and magnitudes, from local lows that bring wind and rain to a small area to huge systems like cyclones that can extend over thousands of miles. Weather under highs tends to be benign, quiet and settled. In the summer, it brings the promise of fine, warm weather. In winter, crisp days, sometimes fog, sometimes cloud – whatever conditions the High sits over, it tends to stick around. High pressure on the surface may look like low pressure's more subdued sibling, but when it continues to build, it becomes resolute. A persistent High can produce persistent weather conditions; if that's fog, air quality can deteriorate, strong sunshine leads to heatwaves and drought.

What's a Blocking Pattern?

Highs or Anticyclones as they are technically called can be transient or persist for some time. In winter, air is more prone to sink when it's cold and dense. Through the winter months, especially over a large land mass, the air can cool down significantly as there is little strong sunlight to warm the surface. This can lead to a quasi-stationary seasonal high during the coldest months. This is why temperatures across Russia and central Europe can fall so low. Scandinavia also can be sometimes dominated by a strong high pressure over the winter season. As the air becomes colder and thus denser, the jet stream bends and curves around it, often weakening and losing definition. The reinforcing of the colder air allows this anticyclone or high to remain in place. This is known as a Blocking High – stubborn regions of air that persist. It's not just winter over land when these highs occur, blocking highs can build and stick around at any time of the year over land or sea. During summer, blocking highs can allow heat to intensify and spread, affecting life on land and water. Heatwaves are detrimental to plants, water supplies and the flora and fauna – pretty much every living thing. During winter, blocking highs allow the cold dense air to cool further and spread, and any moisture turns to ice or snow. The Azores High is a huge semi-permanent high that waxes and wanes over the mid-North Atlantic. Sometimes the Azores High grows and builds and stays. It's part of a family of subtropical highs that punctuate the globe around 30 degrees north. A similar set can be found in the Southern Hemisphere.

It's not just within the blocking highs that extreme weather conditions develop. Like a boulder in a stream, the water flow is disrupted, and everything flows around it. Other times, the boulder is so huge, the flow is totally impeded so low pressures that brew and track are deflected by a blocking high. They skirt the high's edges, and this pattern will repeat as long as the high remains, resulting in a succession of lows delivering unsettled conditions to the same area. The surrounding area, away from the high, is therefore also in a blocking pattern, but this time, it's a different type of weather. If high pressure sits over the UK, low pressure tends to form in the Atlantic and track to the northeast, bringing wet and windy conditions to Iceland. The summer of 2018 was very hot and dry for the UK and much of western Europe due to high pressure that had built up through the spring months. Meanwhile, Iceland had one of its wettest summers on record, as it was in the firing line of depression after depression, as these lows rolled around the periphery of the high, tracking in the same direction.

In the atmosphere there are two types of blocking patterns – the Omega Block and the Diffluent Block. Both result in a persistent weather type, whether it be anti-cyclonic or cyclonic. The Omega Block is named after the Greek letter Omega because it resembles the pattern of the upper-case Omega. The upper air or jet stream flow is amplified, carving out in the upper troposphere the shape of an Omega. High pressure sits under the arch of the Omega shape between a set of fairly stationary lows that make up the two smaller curls either side. The High remains dry with quiet weather, but a deluge of rain and sometimes strong winds persist either side. And if over land, this heavy rain can easily produce flooding and mudslides.

In May 2016, the Omega Block brought extensive flooding to Western Europe with a record rainfall around the Île de France region of Paris and the Seine, flooding parts of the French capital. Similar to an Omega Block is a Diffluent Block, where high pressure sits to the north of low pressure, causing the upper jet stream to split, keeping both parts in place, and ensuring that subsequent dry and wet weather continues.

WHAT IS EL NIÑO?

Every few years the normally dry lands of central and northern Chile and Peru are overwhelmed by heavy rain. When this arid climate takes a temporary U-turn to unusually wet weather locals call it 'El Niño', meaning 'the boy' or, more specifically, the 'Christ child' as it has sometimes coincided with Christmas. Historically, people talked of El Niño as occurring every seven years or so, but the more we learn about this weather event, the more we discover that it rears up regularly and is a far greater entity than originally thought.

The onset of these torrential downpours is a consequence of something that first happens below – in the depths of the neighbouring cold open waters of the Pacific. El Niño begins with the warming of the seas along the Chile and Peruvian coastline. The notoriously cold Humboldt Current that flows from icy climes in the south weakens as a warmer equatorial counter-current pushes south. Normally, the cold seawater keeps the air above cooler and therefore there is less heat energy interacting between the two so fewer clouds form. Although this part of the world gets a lot of fog, rain is scarce and the adjacent Atacama Desert is testament to this – it's one of the driest places in the world. Generally, the soils across this part of Chile and Peru are thin, dry and dusty, lacking vital nutrients to sustain life. Vegetation is sparse yet in contrast the cold seas have nutrient-rich oxygenated water and are therefore teeming with life. Fisheries fuel the local economy, in particular the abundant catch of anchovies.

When occasionally the cold Humboldt Current weakens as a temporary surge of warm waters flood in from the north – in other words, at the onset of El Niño – dark, bulbous clouds gather and then the heavens open. The dry lands can't cope with the sheer volume of water, it is overwhelming. Tricky times follow for the local populations; flooding and subsequent landslides claim livelihoods as well as lives. El Niño may bring much-needed water to the countryside, but the influx of warmer seawater from the equator does more harm than good to the fish population. Whether they wither in the warmer waters or sensibly migrate to colder catchments is up for debate.

Why is This Part of the Pacific So Cold?

A Swedish oceanographer, Vagn Walfrid Ekman, discovered that wind drives the transport of seawater and ice, and not just over the surface layers. This is epitomised in the Southern Hemisphere, where a south to southerly wind will direct the surface ocean current 45 degrees to the left of the wind's path, or in a north-westwards direction, in other words, away from the coastline. This is due to Earth's spin adding an extra resultant force, known as the Coriolis force. However, its influence transcends to some depth of the ocean, directing the mass transport of water to flow up to 90 degrees to the left of the wind direction meaning a huge amount of water flows westwards away from the coastline. The clever dynamics are known as the Ekman Spiral. The converse is true in the Northern Hemisphere, where a prevailing wind will direct the surface current 45 degrees to the right of wind direction, and the mass of water at a depth up to 90 degrees to the right of its path.

The south-easterly trade winds in this part of the world blow northwards, along the eastern coastline of South America. So the theory plays out, not only does the sea current push 45 degrees away from the coastline, a significant depth of water moves away from the coastline at 90 degrees due to wind and the Coriolis force. This allows for cold bottom water from the depths of the ocean to well up and surface. The process is called upwelling and introduces an incredible feed of nutrient-rich, cold and oxygen-abundant water. Hence why the seas here are swarming in marine life, and why the colder water creates a sedated weather environment above.

El Niño has a sister, La Niña (the girl). La Niña is the name given to a significant extra cooling of the waters on the eastern side of the South Pacific, where the water is even colder than normal. Through the decades, the phases of warming or cooling in this part of the ocean have been termed El Niño (warming), Neutral (normal sea temperatures, always cold), La Niña (very cold waters), and many weaker episodes in between. El Niño isn't just a local reversal of ocean currents that interacts with the lower skies and brings storms to the northwest coastline of South America, and it doesn't just affect Peru and part of Chile. It is part of a greater coupled ocean-atmospheric pattern termed

ENSO (El Niño–Southern Oscillation) that stretches across thousands of miles of the tropical South Pacific and its influence can be blamed for many weather anomalies across the world.

The Walker cell is a flow of air that travels on a Business-as-Usual scenario or 'neutral' from the east to west across the South Pacific Ocean. Flowing from high pressure across the southeast Pacific to low pressure across Southeast Asia and northeast Australia. Here, the air rises and travels back to South America, where the cell completes and repeats. Imagine South America on one side and Southeast Asia on the other with the sky above. The circulation could be drawn clockwise, as an upright rectangle – a line extending down to the surface over Peru and Chile, then westwards, over the surface of the ocean, where it rises into the sky over northeast Australia and Southeast Asia. The line then journeys east over the ocean, but this time aloft, connecting to the upper point in the east. Add a rainy symbol over Southeast Asia and a sunny symbol over Peru/Chile – and the mechanism comes to life. Descending air suppresses weather, hence sunshine and cue the Atacama Desert. Rising air allows clouds to form, leading to wet weather – a normal state of affairs over the tropical regions of northeast Australia and Southeast Asia. This is a typical scenario. However, during an El Niño the reverse is true. It rains over Chile and Peru as the air rises (low pressure), and then travels with height westward. Over Southeast Asia and Northwest Australia, the air then descends (high pressure) and the weather becomes dry.

THE BIGGER ATMOSPHERIC PICTURE

This temporary switch in pressure patterns, high to low, at the two ends of the South Pacific is called the Southern Oscillation. Almost like a gigantic see-saw and it's due to the transition to an El Niño event, from a neutral scenario, that resolves back to neutral or sometimes La Niña. This incredible and powerful global weather pattern oscillates over thousands of miles. The combined effect is know as El Niño and the Southern Oscillation, or ENSO. Some years the pendulum swings too far, one way for too long, and the reverse pattern persists. Neither sub-continent can cope with the consequences. For Peru and Chile, the torrential rains bring mudslides, flooding and a dearth of fish. On the other side, SE Asia and NE Australia, the opposite happens – the skies dry up, the rains stop and the verdant, tropical landscape wilts under an unrelenting sun. Drought and crop failure add to the misery as the landscape becomes increasingly vulnerable to wildfires. In the western South Pacific, the sunrays penetrate the delicate shallow seawaters that house a diverse underwater jungle and allows fish and reef to flourish. Too much sunlight bleaches the coral and the abundant submarine life suffers, and sometimes ceases.

Although there are many other factors in play, this huge phenomenon that is so dominant over the South Pacific sends out ripples across other parts of the globe. The Walker cell sweeps to the top of the troposphere, the highest part of the weather machine, allowing the atmosphere in the Northern Hemisphere to be perturbed, even skimming the stratosphere, all of which has impacts on weather cycles beyond this zone.

Even a weak to moderate El Niño can create a trend for colder winters in the southern states of America, with neutral years seeing temperatures on average higher during this winter season. Precipitation rates are seen to drop too. El Niño years can result in fewer cold surges from the Arctic into Canada and the northern states, meaning the northern states are less cold but more likely to have rainfall.

A recent study found that across the Arabian Peninsula, El Niño affected 70% of drought years, especially in the south and southwest. La Niña was linked to 38% of these areas experiencing flood years. El Niño has also been associated with colder and harsher winters in northern Europe. This link is connected with a sudden stratospheric warming that reduces the normally strong Atlantic jet stream, which usually brings milder and wetter weather to the UK and Northwest Europe, and instead easterly winds are allowed to dominate, which in winter means colder spells and more chance of ice and snow.

El Niño has even been blamed for an increased frequency of eastern North Pacific hurricanes and a decrease in tropical Atlantic hurricanes. This is in part due to the upper wind patterns that disturb the vertical wind structure of hurricanes, known as wind shear. So, El Niño can be blamed for a host of weather events across the globe, but it's still not fully understood, so through the year it's monitored closely to anticipate bad weather elsewhere.

On a global scale the El Niño Southern Oscillation can have an impact on the average global temperature. As a strong

El Niño is coupled with a warming of the Pacific Ocean, this heat is transferred into the atmosphere which results in a boost to the global temperature. The warmest years recorded on Earth have coincided with strong El Niño events. Conversely, if a year has been dominated by a La Niña event, when the Pacific waters are cooler than average, the global temperature will tend to coincide with the coolest years.

GLOBAL HEAT AND THE ALBEDO EFFECT

Albedo is the ability of a surface to reflect sunlight with 100% being a total reflective surface and 0% being a completely black surface. Light-coloured surfaces, such as ice or snow, are good reflectors of sunlight and have a high albedo. Dark surfaces, such as green landscapes, absorb more sunlight and reflect far less light. Some of this absorbed light is then re-radiated as infrared energy or heat, therefore darker surfaces are said to have a lower albedo.

The cryosphere is the term given to regions of the world covered by snow, ice or frozen land known as permafrost. This occurs not just across the higher latitudes but at altitude as well. The Arctic is covered in sea ice during the winter months, while the Antarctic is covered with huge sheets of ice over its land mass. Outer sections break off or carve from these ice shelves and float out into open waters, where they eventually melt. The cryosphere plays a vital role in regulating Earth's temperature. On a very basic level, without ice, far more sunlight would be absorbed, land and sea would be a lot warmer and sea levels would be much higher, which would have far-reaching consequences.

In winter, lakes freeze and seasonal snow spreads south. About 68% of all fresh water on Earth is in the form of ice. As a percentage of total water (including ocean and seas) this figure is about 1.7%. This may not seem a lot but its influence on Earth's climate is immense. The yearly distribution of ice and snow varies. The Northern Hemisphere reaches maximum cover during its winter, when the Southern Hemisphere is at a minimum. Ice and snow extends from the South Pole northwards as it enters its winter. At the same time, the sea ice retreats northwards towards the Arctic Circle during the Northern Hemisphere's spring and into its summer. This simplified explanation is neat, and it could be concluded that theoretically, the ice and snow balance across the world doesn't change much through the year. However, there are other factors at play that alter the ice and snow cover through a season and can reduce or increase the albedo, which ultimately has an effect on how hot the planet is. One major contributor to the gradual change in the cryosphere is the rising global heat that in turn is reducing the seasonal growth and enhancing the melting of sea ice. The volume of old ice that is replenished each winter with fresh snow and therefore sustains large glaciers over countries such as Greenland, Antarctica and parts of Iceland is being exposed and also diminishing.

Although solar radiation doesn't change significantly year on year (there is a solar minima every 11 years), the way the atmosphere responds to the Sun's light does vary, depending on the balance of greenhouse gases, large-scale weather events such as El Niño and how much solar radiation is re-emitted as heat across the globe. The most powerful factor in this is air and ocean surface temperatures. Understandably, this is a huge area of ongoing research and particularly in regard to the retreating Arctic sea ice. Many climate scientists project an ice-free Arctic this century. Warming currents can hinder ice growth, angry seas can also reduce the growth of ice sheets during the autumn months. The tipping point is fragile. On a grander scale, a warmer than average winter has a detrimental effect on ice regrowth during autumn/winter. The question is what would happen without the Arctic sea ice?

Without sea ice there would be vast regions of the world that would have darker surfaces; more sunlight would be absorbed, which over time would re-radiate back into the atmosphere as heat. Seas would be warmer, the sea level would rise and there would be far more frequent stormy weather. The boundaries between warm and cold create strong winds, like the jet stream, and are the spawning ground for mid-latitude depressions. Without this difference in temperature the climate of northwest Europe could be a lot different. Many impacts still remain uncertain or unknown.

The rapid long-term melting can be described as a process called a 'positive feedback loop' (or mechanism). In simple terms, this describes the way an initial disturbance on a system, such as the warming of the Arctic, may be enhanced to a greater degree and to a point where there is a higher likelihood that the system will not return to its original state. In this case, with less ice more heat is absorbed, which means the ambient temperature increases, leading to more melting rather than freezing. The next step after a positive feedback loop is the 'tipping point' when a state abruptly changes to another state – for example, the extinction of a species, where there is absolutely no point of return. Many climate scientists warn that melting ice on Earth has such a tipping point; some argue it has already been reached.

Are Glaciers Disappearing?

About 10% of land area on Earth is covered by glaciers. These are frozen masses and store about 75% of the world's freshwater. Glaciers grow and retreat through the seasons, relying on snowfall during the winter months to replenish their volume. Never stationary, they possess their own rhythm. Glaciers are found in many areas of the world, from mountain tops, such as Kilimanjaro in Kenya, to capping volcanoes in Iceland. Valley glaciers form and move through dips towards open water or fjords. Like sea ice, glaciers play an important role in maintaining Earth's albedo. The seasonal glacial melt during the spring and summer services many delicate ecosystems, particularly mountainous biomes.

In April 2019, the scientific journal *Nature* reported a new study had surveyed 19,000 glaciers and found that 10.6 trillion tonnes of ice had been lost between 1961 and 2016 – enough to cover the lower 48 states of the USA in 1.2m of ice. The article suggested 30% of current sea level rise is due to glacial ice melt through these decades. Another independent study published in *The Cryosphere* journal the same year reported 18 out of 19 glacial regions are retreating year on year.

These may be projections based on current and past ice states and how the global temperature is rising, but the evidence of a future with no ice is piling up. Bolivia's only ski resort, located on the Chacaltaya glacier, closed in 2009 because the glacier had retreated so much over 20 years, it eventually disappeared. As ice disappears the world will lose a lot more than just ice.

WHY IS THE STRATOSPHERIC POLAR VORTEX SO IMPORTANT?

The stratosphere is the next layer of the atmosphere above the troposphere. The troposphere is the weather-making zone and the lowest part of the atmosphere in which air cools with height. In contrast, the lowest section of the stratosphere is very cold and instead it warms with height. This is partly due to the presence of ultraviolet radiation and the presence of the ozone layer.

The Stratospheric Polar Vortex (SPV) is a set of very fast winds positioned up to 50km above ground level in the stratosphere. They circulate counter-clockwise across the North Pole during winter (a similar upper-level low pressure system forms across the South Pole in winter in the Southern Hemisphere). The winds develop their strength and direction as air cools during the autumn into the winter. The SPV forms each winter as a result of the large-scale temperature gradients or differences between the mid-latitudes and the pole. This difference is most pronounced in winter, as sunlight leaves the Arctic with no heat source. The vortex continues to strengthen in winter and naturally breaks down when the sunlight returns north during the spring.

Some years, the chain of high-level jet stream winds that propagate around the mid-latitudes can disturb the SPV, weakening and splitting this normally closed winter vortex. These winds sometimes go beyond weakening and in fact, reverse direction. The weaker and reversed winds allow the cold air, normally locked in by the vortex, to sink and compress, and this is when a Sudden Stratospheric Warming (SSW) occurs. The impacts of which can be felt across the mid-latitudes weeks later.

The UK boasts mild winters most years, with a few injections of colder, wintry weather. We call this set-up a mobile scenario; depressions spawn across the Atlantic and track towards the east/northeast, driven by a strong jet stream and carrying milder air from the oceans to the land. During winters when the SPV weakens, or in an extreme episode such as a SSW, the jet stream can lose its strength or zonality. In turn, this can reduce the frequency of weather coming from the west, or the Atlantic – the direction that normally keeps a feed of mild, moist air. This is when the air from the

Continent has supremacy as east/north easterlies, carrying colder air that has lost all its warmth during the continental winter, head towards Britain.

Britain is an island, with hundreds of miles of coastline. The North Sea, despite being a narrow strait of water, is enough to add moisture to these cold east/north-easterly winds. The cold air combines with moisture over the sea to form showers and these readily turn to snow once they hit the colder air over the land. Sometimes this results in a severe snow event across the UK.

What the press dubbed the Beast from the East at the end of February/early March 2018 was an extreme example of how a Sudden Stratospheric Warming (SSW) event a few weeks earlier impacted at the surface level across northern and western Europe. The SSW which happened in January 2018 eventually meant the jet stream lost its mobility and strength; the normally milder westerly winds and associated rain ceased as the wind became easterly. The air got steadily colder across a huge swathe of Europe and high pressure sat over Scandinavia and directed surface winds from the very cold northern continent towards Britain. The dominance of this icy easterly/northeasterly wind resulted in sustained low temperatures across all parts of the country.

Heavy snow showers then penetrated the UK from the east. These developed over the North Sea, lining up in bands and causing widespread chaos and disruption for days. A few days later, a storm that had developed to the south of the

UK, named Storm Emma, tracked northwards and hit the cold air over southern Britain. Rain turned to freezing rain and snow, causing blizzards as gales persisted.

The SSW doesn't always play by the rulebook; it doesn't always translate to a harsh UK and European winter. In 2016, a SSW was in play 50km over the North Pole, but there was no surface evidence because other global circulations reduced the potency of this one event. In January 1987, the UK was blighted by blizzards and sub-zero temperatures for a whole week. This was due to a strong and blocking high pressure over Scandinavia that introduced a blast of Siberian air but no SSW to report.

SSW events do also happen over the South Pole. Their annual occurrence is 4% to the North Pole's 50%. This is mainly due to the more zonal nature of the Southern Hemisphere jet stream. The lack of peaks and troughs mean they don't perturb the Stratospheric Polar Vortex as often. There have only been two SSW events recorded in the past 60 years across the Southern Hemisphere, in 2007 and 2019. When they do happen during late winter there's an increased risk of storminess across Patagonia, New Zealand and Southern Australia but with enhanced westerly winds it becomes drier and warmer over eastern Australia.

Weather patterns are so interconnected that sometimes remote happenings, beyond even the shield of the troposphere, can have fundamental and impactful consequences at ground level and other times the lower level forces direct proceedings at the surface.

WHAT IS THE INDIAN MONSOON?

The word monsoon originates from *Mausam* meaning 'season' in Arabic. It refers to the reversal of wind, from the northeast to the southwest. This flow of air is much warmer and full of moisture, and brings with it a huge amount of rain across the Indian subcontinent.

The heat intensifies over the land through the springtime as the Sun gets stronger. Temperatures across India and Pakistan can soar to the high 40°C and even the low 50°C. The ground becomes parched, aquifers fall dangerously low and rivers reduce to a trickle. The air is dirty and dusty and rainfall is seriously needed. The Southwest Monsoon is a life-transforming weather event for millions. It happens every year around June and within weeks, water has been restored and the oppressive air is cleansed and cooled. It is fuelled by a simple yet powerful mechanism. The switching of wind direction happens because the land at this point in the year is much hotter than the surrounding seas, which is the opposite of the winter months. This significant temperature difference between the ocean waters (20°C) and the scorching heat creates a pressure gradient from south to north, with low pressure over the land mass. As the hot air rises, the cooler, moister ocean air blows in and replaces this ascending heat. The combination of heat and moisture creates huge saturated clouds.

The Sun's progress north of the equator after the spring equinox shifts the global distribution of wind and air pressure. One important zone of weather is called the Intertropical Convergence Zone (ITCZ). Here, global winds over the tropics, loaded with moisture, collide, forming a band of thunderstorms. The ITCZ moves northwards during the Northern Hemisphere's spring/summer and this explains the vast and intense rains that affect the Indian subcontinent.

Like clockwork every year, within a week or so of late May/early June a line of heavy, sometimes thundery and torrential rain forms to the south of mainland India, extending across the Bay of Bengal to Myanmar and Bangladesh. There is such an immense amount of water held in the atmosphere that it bursts to Earth, at times violently. The rains push ever northwards, reaching Pakistan and the foothills of the Himalayas during July. The wind direction and the Sun's strength is most of the story, but the geography of this particular region of the world plays a key role as well. The Himalayas mark the northern extent of the rains. They act as a barrier for any further progression north. Without them, the monsoon would continue its march into Tibet, Afghanistan and Russia. The heat eases and the countryside blossoms into colour again, marking a time of celebration.

For many regions, this six weeks to three months of rain is vital, given how dry it can get during the rest of the year. The monsoon rains restore rivers, returns life to the land and replenishes water reserves for the following eight months. Agriculture relies on this short and wet season. Rice, tea and dairy farms can survive year to year due to the Southwest Monsoon.

Hydro-electric power also relies on a good flow of water so the energy supply and the economy of India and the neighbouring countries hold out for the onset and persistence of the rains each summer. Some have even described the Southwest Monsoon as 'the real finance minister of India'. However, it doesn't take much for this large-scale movement of rain to skew or tip either side of normal. There are many factors in play, and some years, the monsoon can turn from a celebration to a disaster. Too much water, and towns and cities can be submerged, with lives lost and farmland destroyed. The Southwest Monsoon of 2018 may have been unremarkable for many, but in Kerala, western India, the state was inundated with torrential rain, resulting in the worst flooding in 100 years, which led to hundreds of people dying. Too little rain and crops fail, the cost of food and electricity rockets and livelihoods are broken.

Recording-breaking Rain: The Wettest Places in the World

One of the wettest regions in the world is around Darjeeling and Shillong. A lush environment dripping in rivers, forests and waterfalls is perhaps the brighter side of the looming, dark rain clouds that shroud this part of India. The driest months tend to be January and December but even then, 60mm of monthly rain is pretty usual. Here, two places hold the record for the most annual rainfall: Mawsynram and nearby Cherrapunji both average just under 12m of rain a year.

These settlements are located at a high elevation of 1,200m above sea level. The prevailing wind from the Bay of Bengal, laden with moisture, is forced up into these parts from low-lying Bangladesh. The forced uplift cools the air and it condenses out, producing cloud and rain. Elevation is one reason for the weather here, and the other is the monsoon. This part of India is close to the most northerly extent of the Southwest Monsoon's passage, with the Himalayas blocking any further progress. For this reason, the villages are vulnerable to huge amounts of rainfall, which is almost persistent for six months of the year.

There are numerous other heavy seasonal monsoonal climates across the world, all of which are located around the tropics. The intensity, distribution and behaviour of these rainy seasons varies due to the location of land and sea, temperature gradients and associated air pressure responses.

WHY ARE DESERTS IMPORTANT?

Earth has both dry and wet zones; these different climate regions are known as biomes. One-fifth of Earth's land mass is desert. These regions tend to be sparsely populated, with about a sixth of the world's population living in this biome. A desert is defined as a region that receives less than 25cm of precipitation (rain, snow, etc.) each year. But more importantly, it is the rate at which water is lost that creates the desert vista – barren, arid and lacking biodiversity (the opposite of a rainforest biome). It does rain in deserts, although the wet weather tends to be erratic and unreliable, with stretches of months, even years, when there is none. And the evaporation and transpiration rate tend to be high, so any residual surface water is soon gone and most plant life transpire (sweat out) any water content. Evapotranspiration is thus high, with rates exceeding the rainfall rate at a ratio of anything from 2:1 to 33:1, and that's why little vegetation survives here and wildlife is limited.

Where Do Deserts fit into the Global Atmospheric Circulation?

There will always be deserts. The classic atmospheric circulation model describes the global distribution of heat from the tropics to the poles. Although there are yearly variations in detail, it neatly depicts large-scale winds rising and falling as they sweep north and south from the equator. The strongest solar radiation is over the equator, this zone is called the tropics where air rises rapidly to form clouds and rain. The air climbs to the top of the troposphere (the weather-making layer of the atmosphere), spreads out north and south, and eventually descends. Descending air reduces cloud formation, the air is hot and dry, and here, rather than rainforests, deserts preside. This first circulation, closest to the equator, is called the Hadley cell. Two subsequent cells form north and south of this. The Ferrel cell accounts for climatic conditions across the midlatitudes and the Polar cell extends to both poles. As we have seen, there are many deviations from this model, and the atmospheric system is not only more complex, it is constantly interacting with the ocean, the stratosphere above and neighbouring cells. Still, it's a good way to visualise the basic mechanics and offers a simple explanation as to why deserts are positioned on Earth at around 20 to 30 degrees north and south of the equator.

The deserts that sit either side of the tropics are known as subtropical deserts. They are hot places that graduate from the grassy lands of the savannah to arid zones, such as the Sahara Desert in North Africa.

Deserts are also found close to oceans where cold currents run along the coastline. The Namib Desert is located on the western continent of Africa at latitude 24 degrees south. It stretches across the coastal side of Namibia. Its dry climate comes from being positioned next to the Benguela Current, part of the South Atlantic Gyre. This is a cold current that is driven by prevailing south-easterly trade winds from southernmost South Africa and local upwelling close to the shoreline that spews colder water from the depths.

The Atacama Desert also neighbours a cold current, namely the Humboldt, and is affected by local upwelling. This desert is one of the driest place on Earth, with less than 1mm of rain. The other factor in its climate is that the region sits under a rain shadow. It is caught between two mountain ranges: the Andes and the Chilean coastal range. Prevailing winds from either direction mean the air has dried up after its passage across the mountains and descends down the mountainside into the desert.

Although the continent of Antarctica is vast at 13.8 million km² and is known as a land of ice and snow, it is actually classed as a desert. This is because its interior receives less than 51mm or 5cm of ice and snow a year and some parts of this huge land mass are even drier. The McMurdo Dry Valleys are one of the driest and inhospitable places on Earth, so much so that scientists compare this area to the cold, dry desert on Mars. The McMurdo Dry Valleys are close to the Ross Sea. A series of parallel valleys between the Ross Sea and the East Antarctic ice sheet can be picked out easily on satellite imagery due to the dark surfaces and the lack of snow. The valleys sit in the shadow of the mile-high Transantarctic Mountains. Cold and dry air, known as a katabatic wind, with speeds up to 200mph, blows down the mountainsides towards the sea, keeping this region snow- and ice-free.

Not far behind Antarctica in terms of area covered is the Sahara Desert, stretching 9 million km² across North Africa. This desert isn't just a sandy terrain: some parts have sparse vegetation and towards the Mediterranean, plant life extends to woodland and scrub. To the south, the desert land becomes a tropical savannah with seasonal rains that some years allow the landscape to green. The main region is located under an almost permanent high pressure and the stability of the atmosphere makes rain clouds virtually non-existent.

Deserts may not be teeming with life compared to, say, tropical rainforests and are generally harsh environments for anything to survive in, but they do play a key role in adding another level of biodiversity to Earth: species found in deserts are not found elsewhere in the world. This is partly because huge cells of high pressure dominate this region – for example, the Sahara has over 3,600 hours of strong sunshine a year and 86% of daylight hours. The almost guaranteed sunshine is a great opportunity to capture this natural energy source.

What is the Sahara's Broader Reach?

Perhaps an unnoticed characteristic of deserts is that they don't tend to sit in isolation. Their influence, however silent and subtle, reaches far beyond their own borders.

During periods of stormy winds sweeping the Sahara, sand and particles loosen from the top layers of the desert. These are easily lifted through strong winds and the process of convection (the rising of unstable air) and transported at height. Depending on wind direction at altitude, this dust can spread across all parts of the world, but generally, the prevailing easterly winds transport most of it across the Atlantic. Monitoring this by bouncing lasers off plumes of elevated dust as it migrates across oceans from the Sahara has been a key and interesting piece in the puzzle regarding the sustained health and vitality of rainforests. The measurements have captured some 27 million tonnes of dust that eventually reach the Amazon rainforest every year. This dust contains an essential source of phosphorus that replenishes the continual depletion of this element through heavy rains across the Amazon basin. But it's not just the Amazon: evidence suggests close to 43 million tonnes of dust is carried even further, serving the ecological needs of the Caribbean and beyond.

Additionally, a growing body of international research has shown that this dust provides a substantial dose of macro- and micro-nutrients across the rainforest of the Amazon basin. More than half of the world's estimated 10 million species of plants, animals and insects live in the tropical rainforests of the Amazon and one-fifth of the world's freshwater resides there, and this biome incredibly provides 20% of the world's oxygen. So, this essential replenishment of minerals from the Sahara thousands of miles away is a fascinating insight into how Earth's natural systems work in synergy to sustain quite opposite environments.

HOW DO TROPICAL RAINFORESTS INFLUENCE GLOBAL WEATHER?

Tropical rainforests are thick regions of forest that have a rich, extensive biodiversity and a significant amount of rain. This is jungle territory. Only 6% of Earth's land mass may be rainforest, but about half the animal and plant species lives there. Tropical rainforest has an annual precipitation rate between 250 and 450cm, with associated high heat and humidity as well. Over 50% of rainforests fits nicely into the tropical and temperate model, with most deemed tropical. Over 50% of rainforests are found across Latin America and a third of the world's tropical forests are in Brazil. But they are also found in Australia, Asia and a smaller percentage in Africa. In a nutshell, rainforests are the living, beating heart of the global climate. They are a key component of the water table, evaporating out millions of tonnes of water vapour into the atmosphere, cleansing the water system below with nutrients and minerals, and producing and absorbing huge quantities of carbon dioxide and oxygen. Their presence on a global atmospheric scale cannot be underestimated. Rainforests play a vital role in the world's water cycle. Water continually changes state. It can be locked in ice, flowing down rivers, streams and swirling around lakes. Then there's the water vapour that dominates our atmosphere, the most abundant greenhouse gas (clocking up 90% of the entire volume of greenhouse gases). This water vapour can be captured by condensing into clouds that track across the skies, fuelled further by the heat from oceans and land mass below. It's a continuous cycle that cleverly redistributes heat, energy and water across the globe.

Other sources of water vapour, apart from oceans, rivers and lakes, are plants and trees that, through the process of transpiration (exhalation of moisture through the body of the plant) and evaporation (water changing to its gas form) transport cleansed water from the roots into their leaves and ultimately, into the atmosphere as water vapour.

More than 50% (even as much as 75% in dense rainforest) of the precipitation over a rainforest (tropical and temperate) is returned to the atmosphere by evapotranspiration. Interestingly, most of the water released into the atmosphere as water vapour will be returned to the forest as rain – nature's way of guaranteeing that rainforests maintain their water source. The numbers associated with rainforests speak volumes of how important this ecosystem is to the survival of life on Earth. They account for 15–20% of global water evaporation, and more importantly, approximately 65% of the rainfall over land is due to them. Deforestation significantly lowers the levels of atmospheric water vapour through reduced cloud cover and rainfall. The process of evapo-transpiration also has a cooling effect as it takes energy to vaporise water from leaves and keeps the temperature in the forest relatively constant. Reducing forest cover not only depletes soil quality and biodiversity but it also increases the variability of air temperature, with a higher risk of extreme heat during the day. The consequences of depleting the forest floor are far-reaching beyond the boundaries of local ecosystems.

Do Mountains Modify Atmospheric Circulation Patterns?

Any disturbance in the flow of the air will in some way alter its dynamics, whether wind direction and speed and/or moisture content and temperature. The land is accented across many continents by mountains, such as the Himalayas, the Rockies, the Alps and the Andes. These magnificent ranges have a profound influence on regional weather as well as the grander-scale atmospheric circulations. The Andes is one of the longest mountain ranges in the world, stretching 7,000km north–south across South America. It acts as a barrier between the eastern Pacific Ocean and the rest of the continent to the east. It is responsible for the very wet conditions to the east of the range and the arid lands of the Atacama Desert to the west. Also, like other large mountain ranges the presence of ice and snow on its peaks contributes to overall global cooling albedo effect (reflection of sunlight rather than absorption of heat). However, the intense tropical rain that builds up on the eastern side of the Andes fuels the abundant life of the Amazon rainforest, one of the main players in the regulation of carbon dioxide and water vapour across the globe.

The Himalayas play an important part in shaping the climate across much of Asia. This mountain range boasts ten of the world's 8,000-metre peaks. To the south lies the Indian subcontinent, where monsoon rains are concentrated during the summer months. Not only does the position of the Himalayas allow this region to receive a lot of vital water, it blocks the southward penetration of cold Siberian winds, resulting in a warm, tropical climate. But like other large mountainous regions, their presence perturbs and even diverts the larger-scale global winds across the depth and breadth of the troposphere. A strong wind blowing over a peak will send ripples downstream that can trigger weather development. The north–south alignment of the Rocky Mountains that traverse the western side of North America allow wind patterns to massively meander north and south (rather than the general westerly component that dominates Earth's circulation system, due to its spin). Locally, this means a cold southerly can plunge into the southern states of America and peaks in heat can extend into Canada. The cold and dry winds across the Rockies are also a necessary ingredient in the development of tornadoes across 'Tornado Alley'. This vast kink in the wind pattern can and does propagate eastwards across the Atlantic and into Europe. Sometimes a weather development across the Midwest of the USA may eventually affect western Europe in some similar guise, days later (although always modified by the ocean as it pushes eastwards). Like other large geographical features, the position, shape and height of mountains are important factors when describing the atmosphere's bigger picture.

HOW DIFFERENT IS THE WEATHER UP A MOUNTAIN?

Weather within the vicinity of a mountain can be very different: colder, windier, wetter, snow/ice, blizzards, warmer, drier or sunnier. If you get caught up a mountain, the weather can turn at any moment and understanding what happens the higher you go can be life-saving. There are certain conditions when a winter's day can feel like a summer's afternoon on the right side of the mountain.

How Does the Weather Change as it Passes Over a Mountain?

Mountains act as a barrier to wind. They force air upwards and in doing so, its characteristics are modified, cooling is accelerated and condensation increased. Clouds form and then it rains. This type of rain formation is called orographic. This is only half the story: by the time the air has cleared the other side of the mountain it's a different entity altogether. Where you are on the mountain depends on what weather you will experience. In terms of potential differences in weather a mountain can be divided into two parts: the windward side that is exposed to the prevailing wind direction and the leeward side that is far more sheltered from the prevailing winds. Windward weather can sometimes be very different from the other side of the mountain – the leeward side – though not always.

A dominant wind direction has a profound influence on the side of the mountain continually in the firing line. The windward side forces air upwards. The subsequent forced uplift of air cools far more rapidly, producing thick, heavy cloud laden with moisture. Quite often drizzle falls as this cloud, an upslope stratus, clings to the windward side. Sometimes it rains; this is called orographic rain and a reason why this side of the mountain will always be rich in vegetation, despite being often windswept and having many days when the Sun never hits the surface.

It is not only wetter but the higher up the mountain, the colder it gets, and when enough cooling takes place and the air temperature falls below freezing, the rain will turn to snow. This level of the mountain is called the freezing level or snow line. As a storm approaches, and rain turns to snow, the violent winds turn driving rain to blizzards. Zero visibility and freezing air are a lethal combination for any hillwalker. By the time the wind starts its descent down the other side of the mountain, the air has been stripped of its moisture. The sinking of this air as it slides downwards adds the extra forcing of compression. When a gas is compressed, it gains heat, so as the wind flows down the leeside of the mountain, it warms. Crazy! This is known as the Foehn wind or Foehn effect.

The warming and drying up of the air as it flows down a mountainside has other names depending on the region. Over the Rockies, it's known as the Chinook. It's also the mechanism behind the rain shadow effect – the zone that misses out on wet weather. This is evident in the driest places on Earth, such as the Andes, where the prevailing wind has a strong easterly component, leaving the Atacama Desert very dry on its leeward side. Just to illustrate how this can have a profound effect on daytime temperatures, take a 3,000m (10,000ft) mountain, where an air mass that begins its journey on the windward side of the mountain at, say, 18°C, it would eventually transform to a 26°C mass of air at ground level on the leeward side.

Winds close to the surface of Earth are slowed down due to friction. Over the sea there is less friction, yet still the air experiences friction. Above this the air is liberated, it's stronger and freer. This is the gradient wind. As the gradient wind hits the mountain, there is rebound, huge energy and thus, stronger winds. The erratic topography adds extra rapid movement in all directions, leading to gustiness. This is particularly potent when the air is unstable and has a tendency to rise. At the mountain top, wind speeds tend to be two to three times stronger than at ground level.

It's not just unstable air (air that is buoyant and can rise easily) that creates gusty winds. When stable air flows and horizontally hits a mountain, it gets deflected. Rotors of air are then created on the leeside of the mountain. Winds can pick up significantly, leading to momentarily strong surges that cause havoc with low-level aircraft. Evidence of lee gustiness materialises in the formation of rotor clouds aligned with the leeside of the mountain. These rolls of cloud are a result of air condensing as it gusts up into the sky.

Why Is It Colder Up a Mountain?

The rule of thumb is that the air cools by about a degree with every 100 to 150m in height. This is due to a variety of factors:

- Sunlight has the greatest impact on a flat surface (more so than at an angle), so the heat radiating from the surrounding land is warmer.
- With height, the air pressure is lower, allowing the air to expand further, which leads to cooling (the opposite is true when air is compressed, it warms up).

- Winds are stronger aloft so more mixing occurs, which mostly dilutes the heat.

Even though it's colder the further up a mountain you go, you are still more likely to burn at the top than at any other point. Why is this? Ultraviolet rays are stronger with height. On a sunny day, skin burn times are shorter on the top of the mountain than at the bottom. The atmosphere is thinner with altitude, which also allows more UV to penetrate. Conversely, the lowest part of the atmosphere is the thickest, which is good news for the millions living at sea level.

Where Are the Most Extreme Weather Places on Earth?

Deserts, rainforests and mountains are all regions of the world where the weather can be extreme.

As of 2019 here are the global extremes (since records began):

- Hottest place on Earth: 56.7°C, Furnace Creek, Death Valley (Mohave Desert) on 10th July 1913;
- Coldest place on Earth: -89.2°C, Vostok Station (high Antarctic Plateau), 21st July 1983;
- Wettest place on Earth in one year: 26,470mm, Cherrapunji, Meghalaya, India, 1860–61 (rainforest just south of the Himalayas);
- Driest place on Earth: 0mm per year, McMurdo Dry Valleys, Antarctica (cold desert);

- The driest inhabited place on Earth: Arica, Chile (part of the Atacama Desert), with 0.76mm average rainfall per year.

WELCOME TO THE WORLD OF THE JET STREAM

The jet stream is one of the main focuses for any weather forecaster. Analysing the position, shape and strength of the jet stream reveals a huge amount of information about what is happening at ground level. Considering the jet stream is part of a family of upper winds that meanders at great speeds high up in the atmosphere, the connection between surface level and the ceiling of the troposphere may be at first, hard to explain. Yet the jet stream is blamed for everything from an Arctic blast to a heatwave. The big question is why do winds eight miles up in the sky have such a significant effect on the weather at ground level?

The jet stream snakes over the Atlantic Ocean and connects the upper air from North America to Europe. It is part of a larger family of strong upper winds that navigate Earth around the mid-latitudes and are collectively known as Rossby Waves. From a different perspective, looking down at the poles from space, these winds crown the Northern Hemisphere with a waving band of air that sometimes amplifies and at other times is far more zonal, circumventing eastwards. The zone where these strong winds form is where polar air meets tropical air. The official name for the jet stream over the Atlantic is the Polar Jet Stream. Cold air flows south, warm air flows north. These two air masses clash over the mid-latitudes, although due to the undulating nature of the winds, they can extend further north or south.

The core winds of the jet stream tend to be the strongest, especially when the flow is zonal (straight) rather than amplified (peaks and troughs). To the north of these upper winds the air is colder, to the south the air is warmer. This is a good precursor to determine the type of air mass (warm or cold) that may be over a region of land – the cold side or warm side of the jet. The stronger the temperature gradient, the stronger the jet stream, and the greater influence it has over surface conditions.

A stronger temperature gradient also produces large differences in air pressure within this zone. Why? Cold air is heavier than warm air and so exerts a stronger downwards force. The air pressure will fall quickly, with height through a column of cold air relative to warm air. Another way of explaining this is the rate at which pressure changes is swifter in cold air as its mass changes quicker with height because it packs more air into a smaller space and the changes are more rapid.

The boundary between cold air and warm air along the jet stream is pronounced. Although we know that air naturally will flow from more air to less air (high pressure to low pressure), there will never be a direct feed from one to the other because of Earth's spin. The winds on their quest to head into low pressure deflect due to the Coriolis force. This means jet winds tend to flow along temperature gradients, but there will always be the meandering of the wind flow as the air attempts to resolve these gradients.

In winter, the temperature gradient between the poles and the tropics is far greater as the tropics lose little heat through the year, whereas the North Pole is plunged into total darkness and a bitter winter season. Jet stream winds tend to be strongest in winter; similarly, the pressure gradient tends to be strongest then too. All of this has an influence on surface pressure distribution and ultimately, wind, rain and storminess. The shape, position and strength of surface depressions, many of which are formed over open waters, are guided or driven by the jet stream. When the winds are strong, the flow tends to be fairly zonal or straight, flowing west to east. There is less meandering and thus, weather systems track quickly across the Atlantic. The weather set-up is said to be mobile, and if the position of the jet stream is close to a land mass, such as the UK, there can be frequent spells of wet and windy weather.

During the winter of 2013 and 2014, the UK was in the firing line of successive weather systems. The jet stream was overhead. The weather may have been mild, but some parts of the UK had their wettest winter on record, with extensive flooding, most notably across the Somerset Levels; there were also storm surges and wind damage.

When the jet stream is over the UK, the weather is changeable with a higher frequency of wet and windy weather. If the jet stream is to the north of the UK, the country is on the warm side of the jet, which brings warmer/milder conditions. In summer, this produces very warm conditions with most of the rain pushing further north towards Iceland. When the jet stream is to the south of the UK, on the cold side of the jet, temperatures can struggle and in winter this can lead to proper wintry weather – snow, ice and frost.

How Does the Shape of the Jet Stream Affect the Weather?

The shape of the jet stream is key in determining where low pressure and high pressure will be positioned and whether the overall pattern will be mobile or blocked. Imagine a hose of water that is flowing rapidly: as the water is released from the nozzle, it spurts in all directions. The pressure of the water goes from high inside to low outside the hose. A similar mechanism occurs in the jet stream as the air flow diverges out of a core of strong winds. Air is replaced from below, rising rapidly and creating cloud, rain and stronger winds.

At the entrance to the jet stream, air accelerates into it, the mass of air is reduced aloft and this allows low pressure to form at the surface. There are other dynamics at play: columns of air also have a local spin component called vorticity and its strength is determined due to the difference in wind strength and direction with height. With added spin, deeper depressions form at the surface and can result in a higher level of severe weather. The interaction of lower and upper layers of the atmosphere therefore plays out in terms of air pressure distribution and also temperature, which injects further energy into the weather systems.

What are Troughs and Ridges?

Troughs are when the air dives southwards and encourages the development of low-pressure systems. When the jet stream dips southwards and then northwards, making the shape of a U, at the surface air rises rapidly, producing cloud and rain. So the shape of the jet stream is a good indicator of where the surface low pressures will develop. The ridges of the jet stream encourage high pressure, or descending air, producing quieter regions of weather.

When there are many kinks or waves in the jet stream, the whole system propagates slower, like a mature river that flows through the landscape, sometimes taking a straighter path. Where rocks or sediment are stubborn, the flow meanders, other times the flow may completely split. When the flow meanders, it slows and patterns can get blocked. Under a persistent high pressure in summer, heatwaves can hang on, but if the UK sits under a low pressure with nowhere to go because of a blocking high downstream, the rain can keep on falling.

Other parts of the world are also affected by their own jet streams, Japan and North America in particular. A huge chunk of the world's population lives here and the climate is conducive to good living – not too hot most of the time, not too cold most of the time, rain but not heavy tropical rain, a balance of sunshine and cloud, seasons with good vegetation and fertile land. This is thanks to the jet stream bringing changeable weather and moderating the air temperatures by transporting milder air from the oceans during the winter and capping very hot conditions during the build-up of continental heat through the summer – it is only when the broader pressure systems get blocked that extreme weather can ensue.

WHAT IS A WEATHER BOMB?

If you hear the term weather bomb, you might start thinking about governments or organisations trying to harness the power of the weather into a bomb that you could drop somewhere, like bombing a hurricane or tornado on to a place. The idea of capturing the energy created in a hurricane or lightning isn't that far-fetched, but creating a bomb out of it is! So, what is a weather bomb? Its proper meteorological name may give some clues – explosive cyclogenesis.

You will have already heard about tropical cyclones, which are large destructive weather systems that occur in tropical regions of the world; hurricanes and typhoons are just variations of tropical cyclones. You can also get non-tropical cyclones and these are areas of low pressure that form in the mid-latitudes. While these aren't formed in the same way as their tropical cousins, the basic concept of air rotating clockwise (in the Northern Hemisphere) or anti-clockwise (in the Southern Hemisphere) into the centre of an area of low pressure is the same. This inward flow of air at the surface results in convergence, which forces the air to rise into the atmosphere. It's the development and strengthening of these cyclones that we call cyclogenesis.

The Birth of a Storm

We have to start our journey with the polar front jet, which is a fast-moving ribbon of air high up in the atmosphere that meanders around the globe in the mid-latitudes. It travels at 150–200mph at a height of 30–40,000 feet and it's this fast wind that separates cold air originating at the poles, with warmer air originating from the tropics. As it meanders, it is redistributing warm and cold air to different regions. For the UK, the jet's movement around North America and across the Atlantic is of most importance. If the jet stream has big north to south kinks, we start to get small disturbances at the surface, with air either converging or diverging. It is the areas of convergence that we're most interested in when looking at the birth of zones of low pressure. As air converges there is only one place it can go – up. The upward motion of air creates low pressure at the surface and it is in this moment that a cyclonic weather system is created.

If the conditions in the atmosphere are correct, the small area of low pressure formed will continue to grow and develop into a mid-latitude cyclonic weather system. It will come complete with cold, warm and occluded fronts. The whole process of cyclogenesis from birth to maturity normally takes around three to five days. However, if the jet stream is particularly strong and has a sharp enough kink in it, the conditions are ripe for cyclogenesis. Air is moving rapidly up in the atmosphere so pressure at the surface is also falling rapidly. If the pressure falls more than 24 millibars in 24 hours, we call this 'explosive cyclogenesis'. This is also termed as 'bomb genesis' or 'weather bomb'. You can have a mid-latitude storm that's developed from nothing into a violent storm within 24–48 hours.

While explosive cyclogenesis and the term weather bomb have been used in meteorology for decades, it has only really been in recent times that the mainstream media and social media have picked up on the terms. You will often find rather alarming headlines or posts about 'weather bombs' causing end-of-the world style scenarios. In reality, we've always had these types of storms hitting the mid-latitudes and they are often no more than nasty winter storms bringing high waves, strong winds and heavy rain. That's not to say though that we shouldn't be wary of them. As they form so rapidly, computer forecast models sometimes find it tricky to handle the development and tracks of weather bombs. We can often be fairly certain that we'd be in for a spell of stormy weather with the potential for damaging winds but it's important that forecasts are regularly checked as the details of where and when the most severe weather may come could change.

Weather bombs that have hit the UK in recent history include Storm Doris in February 2017 when a woman was killed, with two others seriously injured. Doris brought winds up to 94mph over the high ground of North Wales, with gusts of 60–75mph at low levels. Thousands of homes were left without power and there was major disruption to travel networks across the country.

Another notable weather bomb was the infamous 1987 Great Storm that hit southern Britain. Computer models back then had trouble with the track of the storm and warnings were not really given until it was too late. Wind gusts were up to 100mph, causing the deaths of 18 people, with at least 15 million trees being blown down.

WHERE ARE THE DOLDRUMS?

The Doldrums are a zone over the equatorial ocean where the winds are normally calm but can be punctured by sudden and violent storms as well as unpredictable wind patterns. Five degrees north and south of the equator, where trade winds meet, the Doldrums' lair exists. When sailors spoke of the Doldrums it wasn't the delay in sailing that shot fear into their minds, it was far worse: the Doldrums had a reputation for preventing sailing ships from ever getting to their destination. It wasn't how late they would arrive with their cargo, the consequences of running into the Doldrums reduced the odds for survival. *The Rime of the Ancient Mariner* describes painfully well the tortuous journey through the Doldrums.

Weather at sea in general can be extreme; open waters reduce frictional effects, so winds are stronger. Warm seas around the tropics allow deep air convection to spew up dangerous thunderstorms with hail, lightning and violent winds. The Doldrums are a very different beast. They are a fickle creature that trapped sailors in a zone of calm, with no prevailing winds or strong ocean currents, hindering the passage to distant shores. Air here rises vertically rather than flows horizontally. However, there is more to this zone than benign weather. The tropical air is laden with moisture and decidedly unstable. Storms brew out of nowhere, the atmosphere is energised, ready to spark and offload. Winds can whip up in any direction at any moment as thunderstorms develop, lightning strikes and chaotic gusts do more damage than progress.

The Doldrums are a subset of a larger global feature that circumvents the globe at these tropical latitudes, known as the Intertropical Convergence Zone (ITCZ). To understand the characteristics of the Doldrums, we must delve deeper into the ITCZ. As its name suggests, this is a zone, thousands of miles wide, where air converges. Two sets of trade winds collide – one set blowing from the north towards the equator, another from the south towards the equator. Both are deflected due to the Coriolis force giving them a slight easterly component. Imagine the classic model of three atmospheric cells of air that circulate horizontally and vertically as rising and descending air with surface wind pushing north and south. Where the cells sit next to each other, close to the equator, the winds meet and then rise. It's where these two sets of wind meet and the air converges that the Doldrums exist. The region is identified as a band of thunderstorms, intense heat and moisture, and the air is highly convective. The zone doesn't stay in one place, it moves north of the equator during the Northern Hemisphere's summer, when solar radiation is highest over these parts, and then south when the Sun is overhead to the south of the equator. During the spring/autumn equinox, when the Sun shines directly on the equator, the ITCZ is over the equator. As the ITCZ migrates away from the equator, so the Coriolis force comes into play – Earth's spin starts to have an increasing influence on the air circulation. Localised storms find more momentum and order. Here, deep low-pressure systems develop as the ITCZ combines with the easterly wave disturbances, fuelled by very warm seas below. This is tropical cyclone territory.

The ITCZ across the Indian subcontinent is known as the monsoon trough, triggering huge seasonal rains that push north towards Pakistan, Bangladesh and Myanmar during the Northern Hemisphere's summer. Through the Southern Hemisphere's summer, as the ITCZ follows the Sun southwards, these seasonal rains intensify over northern Australia and Polynesia. Although the Doldrums are quietly present, remaining closer to the equator, tropical rain and storms are never too far away.

DO NORTHERN AND SOUTHERN HEMISPHERES HAVE DIFFERENT CLIMATES?

Global weather and ultimately climate are driven by some simple mechanisms. One being how the surface of the world in its many forms turns light energy into heat energy and then how this heat energy is distributed around the planet. Sunlight has a much more short-term impact on land than at sea. Land mass is more dense and so light waves can only penetrate a matter of centimetres before re-radiating back into the lower atmosphere as infrared energy or heat. When this light energy is lost (during the night or the winter months), the land cools down rapidly and air temperatures can nosedive quickly. The oceans and seas absorb sunlight slowly and the light reaches depths that would never happen on land. This means sea temperatures will always be lower in the summer months than they are on the neighbouring land masses. But the slow release of this heat means that open waters remain warmer during the winter months. This is so different to day and night in a desert; the daytime heat can be unbearable in a 'hot' desert such as the Sahara, but at night under starry skies, there is nothing between the ground and the upper atmosphere to keep the heat in (clouds are good insulators of daytime heat at night) and it gets cold. The diurnal range is therefore broad, night to day, hot and cool. The distribution of the land and sea is significant when it comes to large-scale weather patterns. Let's first of all look at the distribution of land and sea in both hemispheres:

- *Land*: the Northern Hemisphere has more land than the Southern Hemisphere and as the land acts as a heating machine in the summer months, the Northern Hemisphere heats up more. However, continental land masses can get incredibly cold in the winter months.
- *Oceans*: the Northern Hemisphere has less open water than the Southern Hemisphere. This again results in large differences in the distribution of air and ocean

currents, north and south. As previously mentioned, Rossby Waves are one of the main drivers that produce mobile and changeable weather across the mid-latitudes of the Northern Hemisphere. These waves develop due to vast temperature contrasts between the North Pole and the tropics, resulting in a strong band of upper winds that determine and drive weather systems at the surface. Warm ocean currents from the North Atlantic's Gulf Stream and the North Pacific's Kuroshio Current keep things very lively across these parts. They not only moderate air temperatures, but as they push into the northern reaches, the interaction of these warm ocean currents and the atmosphere redistributes heat and moisture across a huge region of the world north to south of the Northern Hemisphere. In direct contrast, the most dominant ocean current in the Southern Hemisphere is the immense Circumpolar Current – a huge mass of water that circumvents the higher latitudes of the Southern Hemisphere. This body of water acts as a gigantic underwater wall, blocking the movement of heat spreading south. Interestingly, the equivalent southern jet stream tends to be consistently stronger throughout the year due to an extreme contrast in temperature from the tropics and the coldest place on Earth, Antarctica. Also, with little land mass to add extra peaks and troughs to the flow, the jet is far more zonal, allowing for stronger winds.

THE BIGGER ATMOSPHERIC PICTURE

Where in the World?
Wettest place: India (Northern Hemisphere)
Driest place: Atacama Desert (Southern Hemisphere)
Coldest place: Antarctica, Vostok (Southern Hemisphere)
Hottest place: Death Valley (Northern Hemisphere)
Sunniest place: Yuma, Arizona (Northern Hemisphere)
Windiest city in the world: Wellington, New Zealand (Southern Hemisphere)

What's in a degree?
70 degrees north: temperate coniferous forests across Scandinavia, Russia, Greenland and Canada
45 degrees north: Mediterranean climate – warm seas, hot summers, mild winters
30 degrees north: mostly dry and hot where major deserts are located
Equator: the tropical region of the world. Hot, humid and full of luscious rainforests
30 degrees south: largely dry encompassing the southern part of Africa to the deserts of Australia
45 degrees south: glaciers and coniferous forests
70 degrees south: inhabitable edge of the Antarctic, home to penguins and icebergs

CYCLONES, HURRICANES AND TORNADOES

You may be familiar from satellite imagery of large storm system that have very distinct centres – the eye of the storm – that, when they track over land, leave a huge trail of destruction. That image could well be a hurricane, cyclone or a typhoon because they all actually look the same. In fact, in essence, they are the same thing – a weather system that can produce torrential rain and sustained wind speeds near the centre of at least 74mph. In meteorology, we call these systems tropical cyclones and as the name suggests, they form in the tropics. They all form in exactly the same way, starting as a tropical depression, strengthening to a tropical storm before potentially becoming a tropical cyclone. They all behave in a similar fashion too, so what's the difference? Well, it all comes down to geography.

Hurricanes are tropical cyclones that are formed in either the Atlantic Ocean or the eastern Pacific. They will tend to impact the Caribbean, Central America and the USA. The official hurricane season runs from June to the end of November, with the most active time being in late August and into September. This is when there is a big temperature difference between the warm ocean surface and the air above. Once a tropical system has been upgraded to an official storm, it is given a name, that has been agreed by a council of the World Meteorological Organisation who are the body responsible for all names given to tropical cyclones. There are six lists of names which rotate every six years and start with A, alternating male and female names up through the alphabet (skipping Q, U, X, Y and Z).

According to the National Hurricane Center in Miami, the average number of named tropical storms is 10, with an average of six becoming hurricanes and two to three turning into major hurricanes (when the maximum sustained wind speed of the hurricane goes above 111mph). The most powerful hurricane ever recorded was Hurricane Patricia, which formed in the eastern Pacific in October 2015. At its most intense, it had a central pressure of 872mbar and a maximum sustained wind speed of 215mph. Hurricane Patricia eventually made landfall in western Mexico, wreaking $460 million of damage.

Typhoons are tropical cyclones that form in the western Pacific Ocean. They have the potential to impact a lot of countries, from Indonesia and the Philippines to China, Taiwan and Japan. Although typhoons can form at any time of the year, they typically develop between May and October with the highest frequency around August. While there are a number of territories around the southeast Pacific that are vulnerable to these storms, it's the northern part of the Philippines (northern and central Luzon) that gets hit the most. Following the same pattern as hurricanes, once a tropical depression in the western Pacific has sustained wind speeds of 39mph and becomes a tropical storm, it is given a name. The Regional Specialized Meteorological Centre in Japan determines the naming of a typhoon but the names themselves are coordinated among the countries that are threatened by the typhoons each year. The most powerful typhoon on record was Super Typhoon Haiyan. It devasted the Philippines in 2013 and over 6,000 people lost their lives. At its most powerful, the sustained wind speed measured 195mph. As is the case with most tropical cyclones, the wind wasn't the biggest killer, it was the storm surge – a huge coastal flood. There was widespread devastation in Tacloban City, with many buildings being washed away and up to 90% of the city destroyed.

Cyclones originate from a tropical cyclone system that has formed in the Indian Ocean and these can impact Australasia in the east to Madagascar in the west. Systems formed in the Bay of Bengal impacting Sri Lanka and India and those formed in the Arabian Sea impacting some Gulf countries are also categorised as cyclones. The Australian Bureau of Meteorology is responsible for tracking and forecasting the paths of cyclones around Australasia and will give the system a name when it becomes a tropical storm. The cyclone season in Australia is from November to April, where there are on average 13 each year. Only half of these become severe and many don't reach land.

Hurricanes and typhoons form in the Northern Hemisphere and spin anti-clockwise due to the Coralis force.

Cyclones that develop over the Indian Ocean also rotate in this direction. But cyclones in the Southern Hemisphere around Australasia, spin clockwise. The most powerful cyclone ever recorded was the 1999 Odisha cyclone (named BOB 06 by the Indian Meteorological Department) in the North Indian Ocean. It was classed as a very severe cyclone with maximum sustained wind speeds of 160mph and brought a 16–20ft storm surge up to 20 miles inland in Odisha, north-east India. The waters completely inundated towns and villages, killing over 10,000 people, with an estimated $4 billion damages bill.

THE FORMATION OF A TROPICAL CYCLONE: THE THREE INGREDIENTS

We've already established that hurricanes, typhoons and cyclones are the same type of storm, and can all be put into the box called 'tropical cyclones'. The biggest of these can cause widespread destruction if they hit land and populated areas so it's very important that forecasters provide enough advance warning if one is expected to make landfall. Understanding the structure, development and lifecycle of tropical cyclones is imperative when giving clear guidance of possible impacts, and this understanding begins with their early genesis.

Let us take you on a journey in the formation of a hurricane. It's September, the peak of the hurricane season, we begin our journey in Somalia. About 3.2km (2 miles) above our heads the African Easterly Jet is flowing from east to west at around 30mph. The heat of the day is producing big thermals and the movement of air in the jet is wobbling, this is our tropical wave – a disturbance in the atmosphere and a precursor to what happens next. This disturbance starts to grow as it moves west across Africa, generating big thunderstorms. *This is our first ingredient.*

We're now in the Ivory Coast and we're getting very wet from torrential, thundery downpours and the flooding risk increases. These thunderstorms are starting to merge together; the wave is now a 'tropical depression' (an organised area of low pressure). The tropical depression, driven by the easterly winds, moves out across the west coast of Africa and begins to cross the open waters of the eastern Atlantic. Two things can now happen. If the sea surface temperature is below 26°C, our tropical depression will start to weaken and die away. However, if the temperature is above 26°C, *we have our second ingredient*. It's this critical temperature that provides the energy needed for the further development of a tropical depression.

Our tropical depression is moving out into the mid-Atlantic, powered by an easterly wind and with a bit of added forcing from the Earth's spin. We now need to look at what the wind is doing in the atmosphere around our tropical depression. In particular, the change in speed and direction with height, known as the wind shear. A strong wind shear will simply rip apart our tropical depression and in fact, weaken a full-grown hurricane. If the wind shear is weak or non-existent then further development can take place. *This is our third ingredient.* As the surface winds increase to over 39mph, the system is promoted to 'tropical storm' status. It's also at this point that it is named.

For our tropical storm to develop further as it tracks westwards, the sea surface temperature must remain above 26°C, and there still needs to be little wind shear within the atmosphere. The other thing that such storms don't like is dust as this can inhibit further growth. The presence of dust and sand, whipped up in the Sahara desert and transported by winds many miles out to sea will disrupt the storm's structure and potential for further development. This suspended dust creates a dry, hot layer called the Saharan Air Layer.

Assuming all of our ingredients are still in place, our tropical storm will continue to intensify and when surface wind speeds peak above 74mph, a fully-fledged hurricane has formed. We then turn to the Saffir-Simpson scale to categorise its severity. Variations of the scale were first used in the 1960s and 1980s to highlight the intensity and severity of a hurricane. The scale takes into account the central pressure, wind speed and potential storm surge of a hurricane, but in its most recent form, we categorise a hurricane only on its maximum sustained wind speed from 1 (the lowest) to 5 (the highest).

CYCLONES, HURRICANES AND TORNADOES

Category	Maximum sustained wind speed	Description
One	74–95mph	Very dangerous winds will produce some damage
Two	96–110mph	Extremely dangerous winds will cause extensive damage
Three	111–128mph	Devastating damage will occur
Four	130–156mph	Catastrophic damage will occur
Five	157mph or higher	Catastrophic damage will occur

Table: The Saffir-Simpson Hurricane Wind Scale from the National Oceanic and Atmospheric Administration. Updated version: 1 February 2012.

WHAT HAPPENS WHEN HURRICANES GO OFF-TRACK?

Most hurricanes begin life as tropical depressions or storms originating around the Cape Verde islands off the coast of West Africa. If the conditions are just right for a tropical storm to develop from here it will travel westwards out into the Atlantic on the prevailing easterly trade winds you find at this latitude. These trade winds will normally continue to push the tropical storm, and the hurricane if it forms, further west along the Atlantic.

Global wind patterns not only produce the infamous trade winds, but also large areas of high and low pressure. Over the eastern Atlantic, a sub-tropical ridge known as the 'Azores high' is a semi-permanent high pressure zone. The winds flow clockwise around a large slack centre, directing storms around its outer realms. In fact, many smaller low pressure systems are caught in the flow on the southern flank of this big high-pressure system that stretches across the Atlantic to Bermuda.

The exact position of the Azores High can have a real influence on the track of storms or hurricanes in the North Atlantic. If the high is weaker and doesn't extend as far west, the hurricane will still travel around the edges of the high but tend to turn and head north much sooner, tracking into the mid-Atlantic with no land to threaten. We call these 'fish storms'. Meteorologists will therefore study the positioning of

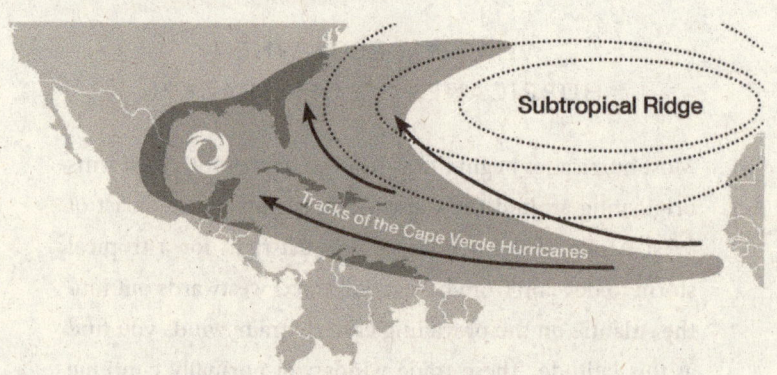

The track of tropical storms and hurricanes around the sub-tropical ridge.

this subtropical high carefully as sometimes, if it persists way out into the mid-Atlantic, it can bring a very active season. Although the subtropical high indicates the general path a storm or hurricane may take, the exact track will depend on other weather and wind systems in play at the time across the Atlantic and even over North America. Warm seas will also have an influence on the storm's direction.

Once storms or hurricanes have started to move north and then northeast around the periphery of the subtropical high, they will normally be into the mid-latitudes and heading back out into the mid-Atlantic. It's through this phase that they start to lose the ingredients and characteristics that made them tropical storms in the first place and become extra-tropical storms. In the mid-Atlantic, these extra-tropical storms get picked up by the south-westerly prevailing winds and the Polar Jet Stream, and head towards Europe. If the conditions are right, the UK may find itself in the path of one of these extra-tropical storms, which bring a spell of stormy weather. In this instance, we would call it ex-Hurricane X, with the X being the name of the original hurricane.

While we've described the movement of storms and hurricanes in a typical situation, the atmosphere is much more complicated in reality. This is one of the beauties of meteorology: no two weather systems or situations are exactly the same.

For a start, not all hurricanes originate from the Cape Verde islands. Some begin life over the Caribbean Sea and others originate close to the waters of Central America as clusters of thunderstorms before moving northwards into the Gulf of Mexico and developing into storms and hurricanes. Depending on the position and strength of the subtropical high, some storms and hurricanes originating from the eastern Atlantic don't even make it out into the mid-Atlantic and may actually travel north, as in the case of Hurricane Ophelia.

The Unusual Case of Hurricane Ophelia

During the hurricane season of October 2017, a small area of low pressure formed over the eastern Atlantic. After a few days of drifting around, it started to show signs of developing into a tropical storm. The National Hurricane Centre monitored proceedings, and despite the sea surface temperature being just marginally favourable for further development, Ophelia became a hurricane. After a few days, despite its location, it intensified into a major hurricane (Category 3) and set the record for the furthest east major hurricane in the satellite era. This was highly unusual and fascinating to watch unfold and there were more surprises to come.

Ophelia became entwined in other weather patterns and pushed northwards, passing close to the coast of Portugal. All the while, it was weakening and turning extra-tropical. Eventually, ex-Hurricane Ophelia made landfall over Ireland and even though it was not technically a hurricane at this point, it had hurricane force wind speeds. Ex-Hurricane Ophelia tracked across the UK before the remnants of the storm drifted towards Norway. One of the other consequences of Ophelia was that the storm system dragged red dust from the Sahara and ash from wildfires over Portugal. As the ash and dust hung in the atmosphere, this extraordinary set of events was captured on camera in the form of an eerie yet stunning red sun as the skies turned orange and red across western Europe.

WHY WOULD YOU FLY INTO A HURRICANE?

Accurately forecasting tropical storms and hurricanes is vitally important for knowing where they are going to impact. Without taking proper action, a hurricane can cause widespread destruction, injury and death. It is also important economically to know when and where a hurricane might hit as damage can cost the government and insurance companies billions of dollars. While we'll cover the monitoring and forecasting of hurricanes in this chapter, the process is very similar for typhoons and cyclones.

Before we construct a weather forecast, an accurate picture of the state of the atmosphere is crucial. There are many methods to record weather observations around the world, from the surface and sky. Since the introduction of satellites in the 1980s, we've gained a far more detailed representation of the atmosphere, as this technology can monitor conditions over the oceans and remote regions of land, places where we can't physically get to on the ground.

Before the introduction of satellites, we had a good network of weather observations from the land but only very limited ones at sea via ships and ocean buoys. The problem with this was that hurricanes form and develop over the ocean, the one place where we simply couldn't measure conditions properly. It was in 1943 when the first manned reconnaissance flight flew into a Category 1 hurricane that threatened Galveston, Texas. The US now have a dedicated team of 'hurricane hunters' who fly out into the Atlantic or eastern Pacific and take meteorological measurements of developing storms and hurricanes.

In the early days, the hurricane hunters would be tasked with flying into storms with an array of onboard instruments that measured wind and atmospheric pressure in all parts of the storm. It was these measurements that would a) determine the intensity of the storm and b) provide data that would go into the forecast computer models to give a projection of its track and development. It is one thing flying into a developing tropical storm that might have wind speeds of 64mph or so around its centre, but these hurricane hunters would also need to fly into the eyes of major hurricanes that had sustained wind speeds of over 157mph, with gusts much higher than that.

You may think that these people are crazy, I mean, why would you want to fly into the most destructive weather system on Earth? Well, those who do it are very humble and pragmatic about it. The key thing for them is to avoid the severe turbulence in the storm – the up and down winds that might give you a bumpy ride at times on a commercial flight. In a hurricane, it's these up and down winds that could potentially break the wings off an airplane. The pilots can track where the severe turbulence is located using onboard instruments and navigate around it. While they will then be flying in wind speeds and gusts in excess of 160mph in the strongest hurricanes, these are straight line winds and something even commercial aircraft are used to handling.

Nevertheless, flying into hurricanes is not completely without danger. Hurricane hunter Jack Parrish from the National Oceanographic Atmospheric Administration (NOAA) told us he grew up with an interest in meteorology, and it seemed like an unreasonable dream to become a hurricane hunter. Similar to loving outer space and having aspirations to become an astronaut. Flying his first hurricane, Allan in 1980, he recalled how on the first day going through the eye numerous times wasn't too dissimilar to flying on a commercial flight. The second day, however, was slightly more traumatic. Severe turbulence meant that most of the instruments within the aircraft came loose and started flying around and they spent most of the day having to clean up.

The hurricane hunters are made up of a combination of National Oceanographic and Atmospheric Administration (NOAA) aircraft and US Navy weather reconnaissance squadron aircraft. Each type of aircraft undertakes different roles in gathering meteorological information, and the crew varies from meteorologists to engineers. Two pilots fly upfront and the flight director coordinates the mission in consultation with the lead meteorologists. The crew monitor the instruments, ensuring the members at the rear of the plane release dropsondes. These are tubes of around a metre long that house weather sensors and a GPS, which are fired out of the bottom of the aircraft, falling through the storm or hurricane, measuring things like pressure, temperature and humidity. The GPS sensor will be able to determine the wind speed and direction. Fifty or more of these dropsondes will be dropped out of the aircraft at pre-planned points of the hurricane to give the scientists real-time readings. When the hurricane hunters go through the eye of the hurricane, the dropsonde they deploy will eventually fall into the ocean and will give the scientists the important central pressure reading.

Another veteran hurricane hunter, Dr Hugh Willoughby, told us about the worst experience he had while flying into Hurricane Hugo in 1989. In severe turbulence around the eye, they lost an engine! With three engines circling inside the relative calm of the eye, they started to take emergency procedures and look for a route out of the hurricane, attempting to avoid more severe turbulence. Thankfully, colleagues in another hurricane hunting plane who were taking measurements above Hugo were able to direct them out of the hurricane to where there was light turbulence.

The introduction of satellites has now helped in measuring, tracking and forecasting these storms but the hurricane hunters still play a vital role. If a hurricane is forecast to come close to making landfall, they will undertake many missions, flying into the storm, taking measurements within it, above it and underneath it. The data gets sent back in real time to forecasters at the National Hurricane Center in Miami to analyse. This up-to-date data will also form the basis of the starting conditions of the hurricane forecast computer models, which will ultimately improve the forecast of the storm's track and intensity.

WHAT IS A STORM SURGE?

When a tropical cyclone hits land, we often think the wind is going to be the biggest impact. However, it's actually the storm surge that results in huge coastal flooding and has one of the most significant impacts. With storm surges in excess of 6–12ft and seawater being pushed many miles inland, it's no wonder that this can be the biggest killer when a cyclone hits.

As the name suggests, we are essentially talking about surges of water associated with storm systems. Whether they are large-scale tropical cyclone or smaller areas of low pressure zones found in the mid-latitudes.

Over water, areas of low pressure will raise the sea level, albeit slightly. For every millibar drop in pressure there can be a 10mm rise in sea level. Disregarding the noise of local sea state or waves, in general, the surface of oceans and seas are not totally level but undulate depending on the distribution of air pressure. In very simple terms, when an area of low pressure makes landfall, it brings a rise in the sea level on the coast, which can result in coastal flooding. This is only part of the process.

Another important consideration is the direction and strength of the wind. The position of the low may direct winds onshore, pushing the bulge of water on to the land. The stronger the wind, the greater the ferocity. A shallow seabed close to the shoreline will also enhance the height of floodwater.

The biggest storm surge in recent times was caused by Hurricane Katrina in 2005, a category 5 major hurricane with winds up to 175mph. The segment of the storm that positioned the winds to blow directly onshore, was along the right flank of the hurricane, and this is where the storm surge was at its greatest. This took place over the western and central part of the Mississippi coastline, causing a 27 ft wall of water to penetrate 6 miles directly inland and up to 12 miles along bays and river. New Orleans was hit hard. The flooding and consequent devastation was caused by the breach of over 50 levees across the city. The flooding and devastation in New Orleans affected 80% of the city where 50 levees were breached. These levees were designed only to withstand a category 3 hurricane.

Big storm surges are not just limited to those associated with tropical cyclones. In the winter of 1953, the UK, Netherlands and Belgium were hit by a big storm surge that flooded a huge area and killed over 2,000 people. This was due to a deep and large low pressure that tracked south across the North Sea and into Europe. This combination of position of the depression as well as high spring tides meant that sea level rose 18ft above normal. In the UK, over 900 miles of coastline was damaged, with floods forcing over 30,000 people to be evacuated from their homes. The most severe flooding occurred in the northern Isles of Scotland down through Aberdeenshire and then further down the coast in Lincolnshire, Norfolk, Suffolk and Essex. The total death toll on land in the UK was around 300 people, with another 223 who died at sea from the storm. It was one of the deadliest natural disasters in the UK. As a result of this, the Thames Barrier was proposed and then eventually opened for use in 1984 to protect Greater London from any future storm surges from the North Sea.

WHY DO WE NAME STORMS?

For centuries, assigning tropical cyclones names has been common practice around the world as a way of identifying storms. Katrina, Patricia and Haiyan are just some of the names already mentioned in this book as infamous and destructive hurricanes and typhoons.

The first time weather systems were given a name goes all the way back to 1887 when an Australian meteorologist, Clement Wragge, started to informally name tropical cyclones from letters of the Greek alphabet (this is still important today) and Greek and Roman mythology, before he retired in 1907. After this, the naming of tropical storms and cyclones had a chequered history, with some groups and organisations unofficially using women's names. In the US, there was eventually some official practice that saw typhoons being named after women, particularly in the United States Army Air Force (USAAF), where it became popular to use the names of their wives and girlfriends in communications about the storms. This initially took off, so much so, the United States Armed Services devised an official list of women's names for Pacific typhoons in the 1945 season. Naming storms wasn't popular everywhere, and the United States Weather Bureau felt it wasn't appropriate to give tropical cyclones a personality. Controversy also continued for many years about the use of female names.

It wasn't really until 1977 when the World Meteorological Organization decided there needed to be more stringent controls and rules regarding the naming of tropical cyclones, so it set up a hurricane committee and took control of the naming of Atlantic hurricanes. It was then that they decided they would use a combination of male and female names following the alphabet. They decided on five lists of names to be used over the subsequent five years. Other centres in the Pacific and Indian Ocean eventually followed suit and today, lists are drawn up and agreed upon ahead of a season.

For the North Atlantic there are six lists of names which rotate every six years. They use the alphabet from A to W, skipping Q and U. If all the names on a list are used, storms are named after the letters of the Greek alphabet. Remarkably, this has only happened once during the 2005 hurricane season. It was the most active season on record, with all the predetermined names on the list having been taken, so the first six letters of the Greek alphabet had to be used: Alpha, Beta, Gamma, Delta, Epsilon and Zeta.

The naming of storms has been proven throughout history to help communicate potential impacts of severe weather. It is thought by humanising weather, storm naming is a consistent way to maintain clear communications from various media and forecast centres, particularly if there is more than one depression occurring at the same time.

In the UK there wasn't an official naming system for storms until 2015. Before that, some major storms were unofficially given names by certain weather forecasting companies that were then picked up by the media. The biggest and most-well known one, especially in the meteorological community, was the St Jude's Day storm in October 2013 – this was named by the Weather Channel UK (their parent company, the Weather Channel in the US regularly started naming winter storms (not hurricanes) in the 2012–13 season). The St Jude's Day storm was one of the most severe wind storms to hit southern England and caused 17 deaths across Europe. As the name was used extensively in the build-up and during the severe weather it was deemed to be a successful method in communicating to the public the forecast.

In 2014, the UK Met Office and Ireland's Met Éireann collaborated to come up with a predetermined list of storm names to be first used during the autumn and winter seasons of 2015 and 2016. The objective was to raise awareness of the dangers of storms and to ensure greater public safety. The names would follow the same World Meteorological Organization format of alternating men and women's names following the alphabet from A to W and omitting Q and U. A storm is named when it is deemed to be capable of having a substantial impact on the UK or Ireland from wind, rain, snow or a combination. The first named storm was Storm Abigail on 10th November 2015, which brought a maximum gust of 84mph to the Outer Hebrides, Scotland, leaving more than 20,000 homes without power. To add a small complication, any ex-tropical storms or hurricanes that track across the Atlantic towards the UK and Ireland will continue to be called the original name allocated by the US National Hurricane Center.

TORNADOS: THE MOST DESTRUCTIVE WEATHER EVENT ON EARTH?

We all have an idea of what a tornado looks like. The whirlwind of air menacingly descending from a big black cloud before touching down and creating huge amounts of damage. On the grand scheme of weather events, tornadoes are quite small-scale but can cause some of the most ferocious winds on the planet. In fact, the strongest wind ever recorded on Earth was from the El Reno, Oklahoma, tornado in 2013, where the wind was recorded at 301 miles per hour.

Thunderstorms are often the precursor to tornado formation so the first thing we need is a big cumulonimbus cloud forming. These clouds stretch many miles up into the atmosphere, with big up and down drafts rotating within the cloud, making them very turbulent. It's these clouds that produce large hail and lightning strikes. While the cumulonimbus cloud is essential, you also need a few more things to occur for a tornado to form. More often than not, you'll get tornadoes forming when these cumulonimbus clouds develop so much that they form what we call daughter cells, essentially other cumulonimbus clouds. Collectively, these are known as 'supercells' and are especially powerful towering thunderstorms. The clouds themselves are actually formed by heat at the ground rising into the atmosphere and in supercells, this rising air is particularly strong and creates huge updrafts into the base of the cloud. Within the cloud, wind speed and direction changes quite rapidly in what we call wind shear. With all these updrafts and wind shear, the air will start to rotate into a column of air known as a vortex at the base of the cloud. This spinning column of air will start to suck up even more warm air from the ground as cold air inside the vortex is forced down. This downward motion will stretch the vortex further down from the cloud, creating a funnel cloud. Eventually, the funnel cloud will grow and stretch towards the earth and only once it touches the ground is a tornado born.

Most tornadoes are quite small, with a diameter of around 80m, travel a few miles and typically, have wind speeds less than 100 miles per hour. However, some of the biggest tornadoes can be around 3km wide, travel up to

100 miles and have wind speeds up to 300 miles per hour. Unlike a tropical cyclone, where we classify the strength by certain wind speeds being met, tornadoes are classified on a combination of the wind speeds measured, along with how much physical damage they create on the ground and categorised using the Enhanced Fujita (EF) Scale.

Enhanced Fujita (EF) Scale		
EF0	65–85mph	Light damage
EF1	86–110mph	Moderate damage
EF2	111–135mph	Considerable damage
EF3	136–165mph	Severe damage
EF4	166–200mph	Devastating damage
EF5	>200mph	Incredible damage

There are a number of different records for tornadoes but the most extreme tornado in recorded history was the Tri-State tornado in 1925. Even though the Enhanced Fujita Scale wasn't in use at the time, it was considered an EF5. It holds the record for the longest path a tornado took, travelling 219 miles through the states of Missouri, Illinois and Indiana (hence tri-state). It lasted for about three and a half hours and travelled at a speed of 73 miles per hour. It was also the single deadliest tornado in US history with 695 fatalities.

The US is known for its tornadoes and one area that has become synonymous with this weather phenomenon is 'Tornado Alley'. Encompassing the Plain States, Tornado

Alley is generally considered to take in Oklahoma, Kansas, the Texas Panhandle, Nebraska, eastern South Dakota and eastern Colorado. This area is particularly prime tornado breeding ground in the spring as it sits in the middle of cold air flowing from the north and warm, moist air that has blown in from the Gulf of Mexico. When these two different air masses meet, the unstable atmosphere is then loaded with moisture and energy to produce the large thunderstorms and supercells.

While you might associate tornadoes with the USA, they can actually form just about anywhere on Earth. It may surprise you to know, but in the ranking of frequency of tornadoes (the number recorded per square mile of land) in any country, the UK sits second on the list behind the Netherlands. The UK has an average of 30 to 50 tornadoes a year, but most residents won't have seen a tornado in the UK, so how does that work?

How Different Are Tornadoes in the UK?

Tornadoes that form in the UK are normally very small and last for a few moments, only inflicting minor damage on property. However, in July 2005, severe thunderstorms broke out across the UK. Supercell development, the precursor to a tornado outbreak, was observed in places across England, with a total of six tornadoes recorded on one day. One of these tornadoes formed in Birmingham and it developed into one of the strongest tornadoes recorded in the UK. The tornado touched down in a built-up area of the city suburbs and travelled around 7km, causing extensive damage to buildings. After the TORnado and storm Research Organisation (TORRO) and Met Office assessed the meteorological data and damage, they estimated the tornado to have wind speeds of between 137 and 186mph. This would have made it an EF2 or EF3 on the Enhanced Fujita wind scale. There were luckily no fatalities, but the cost of the tornado was estimated to be around £40 million, with buildings, cars and trees destroyed. It was the costliest tornado in British history. According to TORRO, the largest tornado outbreak in Britain was in November 1981 when an active cold front swept across Wales and England, spawning 105 tornadoes in one day. The tornadoes were short-lived and weak so didn't cause any fatalities.

In a study of tornadoes across the UK, from 1980 to 2012, it was found that 'Britain's Tornado Alley' was located in an area between London to Bristol and north to Birmingham and Manchester. Here, there is a 6% chance per year of a tornado occurring within 10km of a given location.

Forecasting where tornadoes are going to form in the UK is very tricky. Unlike the USA, where tornadoes mostly form in large supercell thunderstorms that are in the first instance quite easy to forecast and observe, most tornadoes in the UK are created along narrow storms that form along cold fronts. In certain meteorological situations we can forecast when we think these very active cold fronts are going to affect the UK so we may be able to suggest some tornadic activity during this period. We can even be quite

CYCLONES, HURRICANES AND TORNADOES

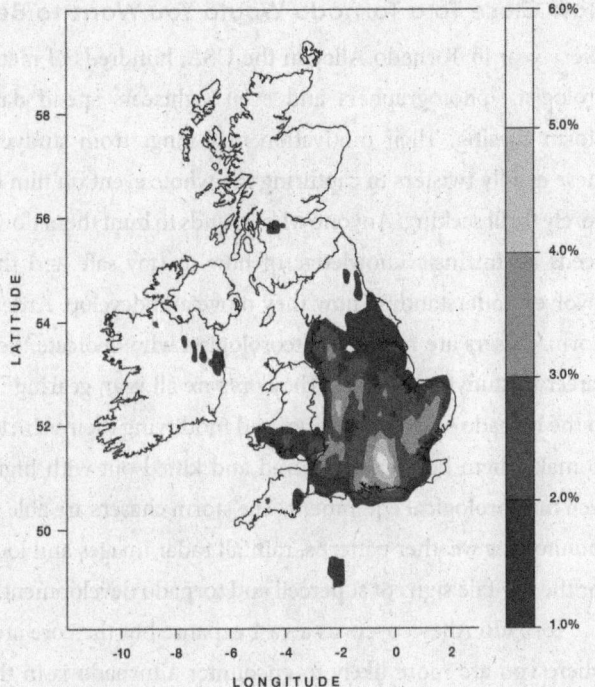

Likelihood of a Tornado forming in a year (%)

From the paper: 'Climatology, Storm Morphologies, and Environments of Tornadoes in the British Isles: 1980–2012', Kelsey J. Mulder and David M. Schultz.

specific on the broad areas that may be at greater risk than others. That's as far as the forecast can go though. Tornado formation is so unpredictable that you can have two days where the meteorological situation looks identical and on one day, you can get a number of tornadoes forming, but on another, none at all.

How Close To a Tornado Would You Want to Be?

Every year in Tornado Alley in the USA, hundreds of meteorologists, photographers and even sightseers spend days 'storm chasing'. Their motivation may range from studying these deadly twisters to capturing the whole event on film or purely thrill seeking. Anyone who intends to hunt them down needs an intrinsic knowledge of how to stay safe and this involves understanding how they move and develop. Ardent storm chasers are trained meteorologists who dedicate their career to studying twisters. They prepare all year, gearing up to the tornado season in spring, and modifying their vehicles to make them heavier, armoured and kitted out with high-tech meteorological equipment. The storm chasers are able to monitor the weather patterns, rainfall radar images and look for the tell-tale signs of supercell and tornado development.

Tornado Alley covers as a vast expanse but the core area where you are more likely to encounter a tornado is in the Dust Bowl Great Plains around Texas, Kansas and Oklahoma. This is partly because the flat open land makes it easier to spot thunderstorm and tornado development many miles away. The downside is that because of the size of the area, it takes a lot of planning and forecasting prior to the chase otherwise countless miles are covered over days without seeing a thing.

For storm chasers, excitement can reach fever pitch during May and June as this is when statistically there are more tornadoes. After selecting an area of potential storm and tornado formation, the drive may take a few hours and atmospheric conditions are continually monitored.

There can be a lot of waiting around for conditions to open and further searching is often required. As tornadoes are anything from a few hundred metres to two miles wide and last a relatively short space of time, chasers can either see very little or find themselves very close to action.

The dangers of getting close to or in the way of a tornado are quite apparent but as it's not just the tornado that occurs in a severe thunderstorm, there are certain risks associated with lightning, large hail and flooding. Often the greatest dangers in storm chasing can come from driving in these conditions. The heavy rain and hail will reduce visibility, with the addition of downed power lines or other debris on the roads that need to be avoided. Once storm chasers have an idea of which storms might spawn tornadoes, they keep a safe distance away, not placing themselves in the predicted path, and have an escape route planned. While there may have been many lucky escapes for storm chasers, with a handful losing their lives due to driving accidents while chasing, for many years there had been no deaths directly inflicted by a tornado until May 2013.

The Record-Breaking El Reno Tornado

It was during the infamous El Reno, Oklahoma tornado, the widest tornado ever recorded, when three storm chasers were caught in its rapid development and were killed. The atmosphere was primed that day with a lot of moisture and warmth so storm chasers knew there would be some big storms developing. As the day progressed, storms fired up with a supercell developing quickly and a tornado touching down soon after. Most tornadoes travel in an easterly direction, but this one started to move in a south-easterly direction, which meant the tornado was heading towards the areas where people were observing the tornado. The three storm chasers killed that day were highly respected professionals with years of experience, involved in the research of tornadoes under a project called TWISTEX. The team were attempting to take observations of the powerful EF3 tornado when they realised they were too close and attempted to escape. But as the funnel cloud rapidly expanded, it enveloped the car and threw it approximately half a mile from where they were. The deaths shook the chasing community and even though it was rare for there to be any fatalities, it brought home how unpredictably tornadoes can behave and the risks involved when people get too close.

WHAT WEATHER IS MOST LIKELY TO KILL ME?

Before we create mass panic and you start worrying about being struck down by the weather, let's just begin with one big caveat: it all depends on where you live! Some weather events only affect certain countries and where you live in a particular country will also make a difference. For example, the UK cannot be hit by a hurricane so that's not going to kill you there. Or if you live in a mountainous area, a coastal storm surge isn't going to kill you. There are many different types of extreme weather we can get around the world: tropical cyclones, heatwaves, drought, floods, very cold weather and thunderstorms. While floods and tropical storms may affect thousands of people at one time, they are not as frequent as, say, thunderstorms, which happen all over the world regularly but may only affect fewer people at any time.

Let's start with a thunderstorm. A lightning strike coming out of a cloud has around 1 billion volts of electricity and is about five times hotter than the Sun. Lightning strikes happen more often than you might think, with around 100 strikes hitting somewhere on Earth every single second. The vast majority of these obviously go unnoticed or hit buildings that have lightning conductors, which transfer the electricity to Earth. Lightning can strike people and it's estimated that around 4,000 people worldwide are killed by lightning every year (some suggest the figure could be as high as 24,000), with many more being hit and injured. The documenting of lightning fatalities in every country is sparse, but the available data gathered for the International Lightning Detection Conference suggests that India has the highest number of fatalities, with around 2,000 people a year. But per population rate, you're more likely to be killed in Malawi, with an average of 45 people being struck dead there every year. For reference, the figure is just two people a year on average that are likely to be killed by a lightning strike in the UK – so, if you live there, that makes your chances about one in 70 million.

Heatwaves and cold weather events can cause deaths to a larger proportion of people at any time. According to the United Nations Office for Disaster Risk Reduction (UNISDR), extreme temperatures caused 27% of all deaths attributed to weather-related disasters between 1995 and 2015. In this time, 164,000 people were killed due to heatwaves. During a heatwave the body can't cope with the higher than normal temperatures and can start to shut down unless precautions are made to cool ourselves down. In 2003, there was a Europe-wide heatwave in which temperatures were above average, with records being broken in many countries, and more than 72,000 people lost their lives due to the excessive heat. Similarly, in Russia in 2010, a heatwave caused the deaths of more than 55,000 people.

In 2003, there was a Europe-wide heatwave in which numerous temperature records were broken in many countries. However, excess winter deaths due to colder than average weather kill many more. In a typical year there can be 25–30,000 excess deaths. Many of these are people who are over 85 years old, for whom the cold weather can cause respiratory problems such as bronchitis, asthma and flu. These deaths are mainly attributed to fuel poverty, which means the low-income older population are unable to adequately heat their homes.

Worldwide, floods and droughts can affect billions of people every year, especially in developing countries. It is very difficult to put exact figures on the number of people who die during a flood or drought as reporting is very sparse. The UNISDR report mentioned previously suggests that Asia and Africa are affected by flood more than other continents, with the estimate on flooding deaths being around 8,000 people per year worldwide.

The biggest weather killer worldwide though is storms, including hurricanes, cyclones and their associated storm surges. It has been estimated that between 1995 and 2015 more than 242,000 people were killed by the impact of a storm. The sheer size of a storm as it makes landfall means its impacts are felt over a widespread area. The wind will cause severe destruction, with the rain and storm surge bringing widespread flooding. In fact, as we have explored, the flooding during a storm is more likely to kill you than the strong winds. As you might imagine, the impacts on a developing country are far greater than those on a developed country, where the infrastructure and emergency responses will be superior.

WEATHER PHENOMENA

The drama that continually plays out above us has a physical effect on our daily lives. Occasionally we look skyward and take a breath. Something up there has momentarily made time stand still. When sunlight meets air, the sky can morph into a theatrical display of many shapes, patterns and colour. Sometimes magnificent clouds don't foretell formidable weather but seem to hint at the mark of an artist's paintbrush. This is weather in its most stunning artform. From time to time we are simply amazed by what's happening above us and yet our quest to understand why never wanes ...

RAINBOWS

How Are Rainbows Formed?

I'm sure it's not just us weather geeks who still get amazed when we see a rainbow in the sky. Once one appears, you're searching for the pot of gold at the bottom of it or reciting 'Richard Of York Gave Battle in Vain' in your head! The word 'rainbow' comes from the Latin *arcus pluvius*, which means 'rainy arch'. They have been spotted throughout history and are often part of mythology. They even appear in the biblical

Book of Genesis, Chapter 9, as part of the story of Noah and the Great Flood. According to the Bible, after the flood, God created a rainbow as a sign that he would never destroy all life on Earth with a flood again.

Rainbows are a sure sign that simultaneously it's raining and the Sun is also out, and this helps us explain how they are formed. Rainbows don't actually exist, i.e. they are not a physical thing, they are an optical-phenomena based on the interaction between sunshine, water droplets and your eye. Because of this last element, your eye, every rainbow is completely individual to you. Someone standing only metres away may well be seeing a rainbow that is ever so slightly different to the one you are seeing, all because of the different angles at which light reaches the back of your retina.

You need sunshine and water droplets for a rainbow but it's not that simple – it's also important to have the Sun shining behind you and the water droplets in front. To explain how we see the wonderful arc of colours, we need to touch on a bit of physics. When sunlight hits a spherical water droplet, part of the light is reflected back but the rest of it goes inside the drop and is bent into a different angle, known as refraction. Some of this light then hits the back of the drop, where it is internally reflected. The angle of this reflection is around 42 degrees but because colours reflect at tiny differences from each other, the white light from the Sun is dispersed and split into the different colours of the light spectrum, i.e. violet light has a shorter wavelength and is reflected at a greater angle than the red light.

As the light travels back towards your eye, you will see the full spectrum of colours from red to violet. While we often separate the colours into bands of red, orange, yellow, green, blue, indigo and violet in a rainbow, in reality they all blur into each other. The red part of the rainbow is always on the outside of the arc, with the violet on the inside, and even if you don't see the full spectrum of colours between, this is always the order. Except where you have a secondary rainbow – the double.

How Do Double Rainbows Work?
In theory, all rainbows will have a double as the sunlight is often reflected twice within a raindrop. In practice, because light escapes during the second reflection, it is often fainter and the colours are less distinct. The second rainbow is also bigger and spread over a greater part of the sky. When the light is reflected for a second time, the colour sequence is actually reversed so on the second rainbow, the outer colour will be violet followed by indigo, blue, green, yellow, orange and red on the inside.

You might also notice the sky between the two rainbows is darker, essentially because there is little light left to get to your eye within this band. This is interestingly known as 'Alexander's band', after Alexander of Aphrodisias, who first described it in 200 AD.

Reflected Rainbows and Moonbows
Another optical phenomenon you may come across is a reflected rainbow. This almost looks like a double rainbow,

but the two arcs don't follow the same curve. Or you may even see a third bow in between a double rainbow. Reflection bows are most often seen when you have a body of water such as the sea or a lake nearby, but they are very rare. Sunlight reflects off the surface of the body of water and then through the raindrops at a different angle to the direct beam of sunlight that creates the main rainbow. The reflected bow often starts at the same point on the horizon, but the centre of the arc is higher in the sky creating a different curve to the original.

Moonbows are simply produced by moonlight rather than sunlight and other than this change in light source, they form in exactly the same way. Moonbows will appear fainter with colours less distinct, or may even seem white. This is because the amount of light coming off the Moon is less than sunlight, so it doesn't excite the colour receptors of our eye. Interestingly, if you were to photograph a moonbow with a long exposure, it would pick up the colours.

Of course, not all rainbows have to appear in the sky after a bit of rain. As long as you have your back to the Sun, there are water droplets in front of you and the angles are right, you can get bows in waterfalls, mist or even when using a hosepipe to water the garden!

Rainbows in the Clouds?

You may have seen a circumzenithal arc in the sky but never realised what it was you were looking at. In fact, they are often spotted and reported to us as 'upside down' rainbows or grins in the sky as the spectrum of colours reaches up into the sky in an arc rather than down to the ground. Unlike a rainbow, where the spectrum of colours is formed by the refraction of water droplets, they are a type of optical phenomenon that require ice crystals in the sky. Most of the time, you'll see a circumzenithal arc in a cirrus or cirrostratus cloud, which happens to be made up of ice crystals. Because this colourful arc is formed from sunlight being reflected and refracted in ice crystals, it belongs to a family of different optical phenomena called halos. You may also get to see a full 360-degree circle around the Sun called a 22-degree halo or even catch a glimpse of *parhelia* (known as sundogs or mock suns), which appear at the three and nine o'clock positions from the Sun.

To understand how circumzenithal arcs form, we need to do another bit of physics. The structure of an ice crystal can be quite complicated and is dependent on the temperature and pressure of its surrounding environment. They all have one thing in common, though: they have two faces and six sides, making them hexagonal prisms. Circumzenithal arcs are formed when this hexagonal prism is oriented horizontally and stacked up like plates. Light from the Sun enters the plate from the top flat plate and exits from the side vertical part of the prism. The angle (90 degrees) of this refraction within the crystal separates the white light into the different colours you see. It's most likely you would have done an experiment just like this in your school science lessons when you shine a light through a prism on a white bit of paper and see colours coming out the other side.

To spot a circumzenithal arc, cirrus or cirrostratus clouds must be present high up in the sky. It is often the type of cloud that makes the sunshine turn hazy/milky. When the sun is at a low angle you may notice the arc above that at the 12 o'clock position. The centre of the bow is always towards the Sun, with the red colour on the outside going through the spectrum of colours to blue/indigo on the inside.

Do Sundogs Bark?

In almost identical conditions where there is a lot of the ice crystal-based cloud of cirrostratus in the sky, you may get the chance to see a 22-degree halo or sundog. The halo looks like a large disc around the Sun but, unlike the circumzenithal arc, where you get quite distinct colours, the halo appears more washed out, with perhaps just a tinge of red visible. It is formed again from the refraction of sunlight through the hexagonal prisms of ice crystals. The orientation of the prism as light passes through it is different in halos to arcs. In fact, most of the rays are deflected through angles near to 22 degrees to form the bright inner edge of the halo, hence its name.

Halos aren't just a daytime optical phenomena. If the Moon is bright enough in the sky (most likely, a full Moon) with a veil of cirrostratus cloud, the same physics can occur

Sundogs

and you might be able to ascertain a faint 22-degree halo circling the Moon.

If there is a halo (or part of one) present in the sky, then you might also see the sundogs at the three and nine o'clock positions from the Sun. Essentially brighter parts of the halo at these positions, they are most often found when the Sun is low in the sky and while they appear bright, you may also see the spectrum of colours as you would in a circumzenithal arc. If they are really bright, you'll understand why they are sometimes called 'mock suns' as it appears there are three suns in the sky.

CLOUDS OR ALIEN INVASIONS?

The atmosphere is such a dynamic space that you can witness different cloudscapes every day. The clouds are constantly evolving shape as they develop, decay and move in the winds above our heads. Many of us will recognise the regular clouds in the sky, whether it be the fluffy cumulus cloud that populate our skies during summer months or the constant grey stratus or stratocumulus cloud that are more common in winter. Sometimes though you may get to see something extra special in the sky.

Flying Saucer Clouds

Altocumulus lenticularis are a type of cloud that look like lenses or flying saucers in the sky, and don't move with the high-level winds. They are most commonly found to the lee of high ground, formed by the flow of air over the top of a mountain. As moist and stable air doesn't want to go up or down, when forced up the mountainside, its properties of humidity and pressure are being put into a state that it doesn't like, and it therefore wants to go back to the height it was at before and creates what we call a 'standing wave' downwind of the mountain. If the temperature at the crest of the wave is at the dew point temperature of the air, it will condense and form cloud. As it takes quite precise conditions for the lenticular cloud to form, it will very often be elongated, have smooth edges and look like a lens. With the wave motion of the air being generated by a hill or mountain carrying on for many miles after it started, you may get the condensation into lenticular cloud some distances away from the mountain. There can also be a number of lenticular clouds forming, some even at different heights that sit on top of each other. Because of their smooth lens-like appearance, they are often reported as UFOs or 'cloud cover' around a UFO to disguise it. In the US in particular, there have been

Altocumulus Lenticularis

many such 'sightings' and generally the authorities conclude that they are actually these altocumulus lenticularis clouds. Or are they?

Hole Punch Clouds

This is another one of those clouds that on first sight may make you think there is some sort of supernatural invasion taking place! As the title suggests, this phenomenon looks like someone or something has used a hole punch and made a hole in a sheet of cloud. The technical name of this cloud is a fallstreak hole. In simple terms, it is created when part of the cloud layer forms ice crystals that are large enough to fall out of the cloud as a fallstreak. The interesting thing is how we suddenly get ice crystals forming in a seemingly random part of a cloud.

At this point, it is worth remembering that many clouds contain supercooled water in which the temperature of water droplets is actually below freezing. Within a cloud

Hole Punch Clouds

layer of supercooled water droplets, cooling can suddenly occur when an aircraft passes through it, particularly over the wings or through the propeller. This change can be big enough to start the freezing of the water droplets. When the ice crystals start to form at one point in the cloud, a domino effect called the Bergeron process sets off, whereby surrounding water droplets also turn to ice. The heavier ice crystals then start to fall out of the bottom of the cloud. This explains why you've got this streak of cloud beneath the hole, but doesn't fully explain the hole itself. As the droplets turn from liquid water to solid ice during the Bergeron process, a tiny amount of heat is given off. This makes the air expand and rise slightly, which in turn causes the surrounding air to sink. The sinking air around the fallstreak will then warm before the water droplets evaporate, creating a clear bit of the cloud in either a circle or ellipse shape.

WEATHER PHENOMENA

THE RAREST CLOUDS

Some clouds are much rarer to spot than others and in this section we'll highlight three clouds that you'll need to be very lucky to spot yourself along with some other cloud based phenomena. They require just the right conditions coming together to form, which is why they are so unusual. In our opinion, the rarity of these clouds makes them even more special and beautiful to witness.

Udders

Mammatus is a very distinctive cloud that takes the appearance of bulges underneath or to the side of a menacing cumulonimbus thunderstorm clouds. Its Latin name, *mammatus*, is translated as 'udder' or 'breast'. Once you spot one, you'll see why. They only form with cumulonimbus cloud

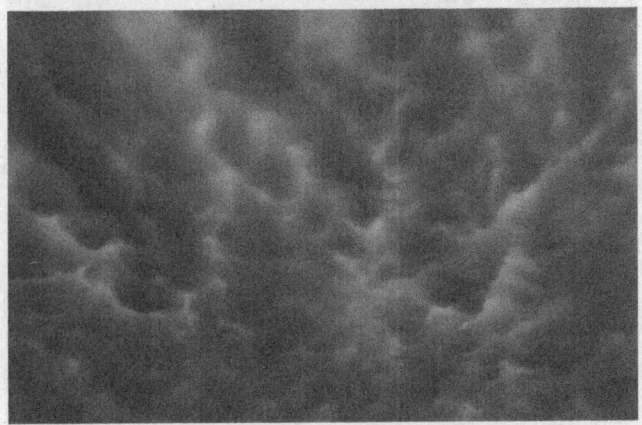

Mammatus (Udders)

as this has the most violent up- and downdrafts needed. As we now know, normal cloud formation is caused as air rises, cools and condenses, creating the water droplets you see as cloud. However, because of the severe downward motion of air in a cumulonimbus cloud, the air and moisture are forced to protrude out of the bottom of the cloud and it is this that forms the bulges or pouches of cloud associated with mammatus.

If you notice mammatus cloud have formed on one of these cumulonimbus clouds it means that the air is extremely unstable and you are likely to get some strong winds, torrential rain and even hail, along with the lightning and thunder. You may want to gaze at its beauty, but you can be sure you're going to need to take cover very soon!

Mother of Pearl

Most clouds form from ground level (also known as fog) to as high as 20km (35,000ft); this is the troposphere. There is a rare cloud, that can form higher than that in the stratosphere. These are technically called Polar Stratospheric Clouds (PSCs), but are most commonly known as Nacreous or Mother of Pearl clouds. As the official name suggests, these clouds form in the stratosphere (around 50,000ft and higher), where it is normally too dry to produce cloud and are most likely found above the polar regions, though if the upper atmosphere is cold enough at lower latitudes during winter, they can also be seen in the UK. On rare occasions when the upper atmosphere is cold enough, even at lower latitudes during winter, they can be spotted. The image of these vivid iridescent clouds made the headlines in the UK during January 2017, photographs captured them over Cumbria. It was suggested that bitterly cold conditions over the stratosphere in the Arctic has drifted south, causing this cloud phenomena to form further south. In this case, like most, these colours really come to life around twilight, when the sun is just below the horizon. Like the altocumulus lenticularis, this cloud is a wave cloud but made up of very cold ice crystals (around -80°C). To get that beautiful iridescent colouring, the nacreous cloud has to be thin and have very small and similar-sized ice crystals in it. Sunlight passes through them when the light is diffracted (when light bends around objects) and scattered into the spectrum of colours. You can see a similar effect on the surface of soapy bubbles.

Kelvin-Helmholtz Cloud

These are perhaps one of the rarest cloud formations to see. In recognition of them looking like breaking waves crashing on to a beach, they are known as Kelvin-Helmholtz waves and are named after Lord Kelvin and Hermann von Helmholtz, who both studied the physics of atmospheric instability. There's a clue there in how these breaking waves are formed – instability. When two layers of air in the atmosphere move at different speeds, there is a velocity difference between the two fluids. This difference creates

Kelvin-Helmholtz Clouds

instability such that the faster-moving air can 'pick' up the top of a cloud layer into these wave rolling structures. When this happens, you may see a whole line of the waves breaking in the sky.

The same process actually happens at the sea surface. The wind blowing over a slower-moving body of water creates the Kelvin-Helmholtz instability which leads to wave formation. If conditions are right for Kelvin-Helmholtz clouds in the atmosphere they are easy to spot as their unusual shape is very different from any other cloud. The base of the cloud will be horizontal with the crashing clouds on top. They don't usually last that long so you'd need to get your camera/phone out quickly to take a snap of them.

Crepuscular Rays

Shafts of sunlight as they pass through a cloud and touch the Earth's surface can add a stunning effect to the cloudscape.

They occur more often during first or last light, when there's an orange glow to the sky.

As you get these shafts of light and shadows breaking through parts of a cloud, some often call these 'God rays' or 'Fingers of God' as some kind of sign from heaven. In reality, it's another meteorological optical illusion as while the shafts of sunlight appear to converge to a point beyond the cloud, they are actually parallel. The apparent convergence is the same optical illusion you get when looking down a straight road or rail track, which appears to get wider nearer to you than the horizon. The scattering of light from air molecules, dust and water droplets also adds to the illusion. The particles in the air scatter more of the short wavelength colours of blue and leave you with the yellow or orange side of the spectrum of colours.

Crepuscular Rays

Anti-Crepuscular Rays

Slightly harder to spot but just as interesting are anticrepuscular rays, which occur at the same time but on the opposite side of the horizon to the Sun – known as the anti-solar point. This time the shafts of sunlight appear to converge on the horizon rather than behind the cloud to the Sun. Because of the lack of light, the rays are much fainter and therefore harder to spot.

Anti-Crepuscular Rays

WHAT'S A WILLY-WILLY?

You might be thinking why is there a section in this book called willy-willy, and what has that got to do with the weather? Well, we're teasing you slightly here because this is the name Australians give to a dust devil. Dust devils are actually known as many things around the world, but willy-willy came from Aboriginal myth in which a 'willy-willy' represents spiritual forms.

What are they? They almost look like a tornado but crucially, they're not and don't have anywhere near the destructive force of a tornado (though some can be powerful). The similarity with a tornado is purely that they are both weather phenomena that have a vertically oriented column of wind. But while you need a supercell thunderstorm for a tornado to emerge from the bottom of a cloud, dust devils seemingly appear without any cloud and form from the ground up. They are visible due to dust being picked up by the swirling wind and transported into the sky.

How Do Dust Devils Form?

One of the key ingredients for a dust devil is… dust. Unsurprisingly, they most commonly form in desert or semi-arid areas where the ground is very dry. On a hot day, the dry ground can get very warm and strong updrafts will start to

Dust Devil

form. As the air rises, it starts to rotate around a vertical axis and with further height the air begins to be stretched. Like a figure skater pulling their arms in to spin faster, the spinning air intensifies. This 'vortex' of spinning air carries dust; this is a dust devil. As long as hot air continues to rush into the bottom of the spinning funnel, they will persist and sometimes intensify. The hot air will cool with height, falling back to Earth and thus maintaining the familiar funnel shape. Inevitably, the availability of hot air is cut off and as cooler air starts to get sucked into the vortex, it will start to dissipate quite quickly.

Dust devils are typically only around a few feet in diameter and rise a few hundred feet into the sky. They might have wind speeds of around 40–50mph so don't really cause any damage. However, they can on occasion be bigger and grow to about a few hundred feet in diameter and rise to 1,000 feet in the air. On these rare occasions, the wind speed can be as much as 60–75mph and lasts for a longer period of time. When this happens, there can be some significant damage to structures and some dust devils may even cause injury.

Snow Devils

You don't need to be a meteorologist to work out that a snow devil or 'snownado', as they are sometimes called, looks very much like a dust devil, but with snow. We should mention here that even though it takes part of its name from a tornado, it is not classifiable as a tornado. It's another spinning vortex of air and snow extending from the ground.

These are extremely rare, with only a small number of sightings caught on camera. The meteorological process involved in its formation is slightly more complicated than that of a dust devil and you need very precise conditions for it to come together.

The spinning vortex is still created by updrafts near the ground surface but unlike in a dust devil when the driving force of the updraft is caused by air at the surface being heated and rising, the driving force is a column of colder air above the ground. This means a temperature difference is created with relatively warm air at the surface and colder air above, forcing air at the surface to rise. But to get a snow devil, you'll then need some kind of wind shear in the lower atmosphere, where the wind speed or direction changes with height. This will generate the spinning air and with lifted powder snow, it will make the spinning column of air visible. Snow devils will only last a very short period of time and, like dust devils, are most unlikely to cause any damage or injury.

Fire Devils

Guess what? If one of these spinning columns of air happens in a fire, you could see a fire devil or 'firenado' as they are commonly known. These are very dramatic vortexes of ash and fire, which add to the ferocity of a raging inferno. In this instance, it is the intense heat of a fire at the surface, where the temperature can reach over 1,000°C, that creates violent updrafts. As more heat is generated at the surface than in the case of the dust devil, a fire devil normally has a tighter and more vigorous vortex extending higher into the sky. They are most likely to occur during wildfires when there is the presence of a strong wind. This, combined with a firestorm creating its own strong wind, can generate a lot of wind shear that aids the formation of the vortex.

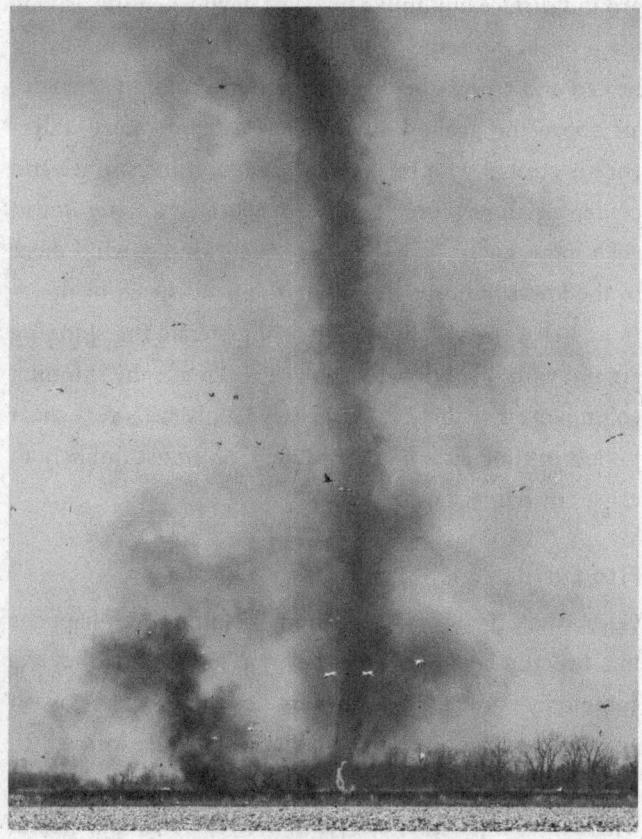

Firenado

THE NORTHERN LIGHTS

You may have been lucky enough to see the Northern Lights in person. For many of us it may be one of those things on the bucket list – it's certainly on Simon's! (Clare has been fortunate enough to see it for herself.) The array of green, purple and sometimes red dancing in the night sky looks so magical.

The proper name for the Northern Lights is the Aurora Borealis and you find it predominantly around the Arctic Circle. There is an equivalent in the Southern Hemisphere, called the Aurora Australis. Technically, there is no difference between the Borealis and Australis, but you tend to find the Borealis is more popular. This is down to the fact that there is more land around the Arctic Circle, so there is greater opportunity to visit a populated area where you might actually see the Northern Lights. Around the Antarctic Circle, you've pretty much got the Southern Ocean so unless the Southern Lights extend northwards into New Zealand, Argentina or the Falklands, you won't have many opportunities to spot them.

What Exactly is the Aurora?

The aurora is caused by the interaction of charged particles, comprised of electrons from the Sun, with Earth's magnetic field and atmosphere. Earth's magnetic field is really complex but put simply, we know that it behaves very similarly to a basic bar magnet aligned north to south, where an electric current flows and extends out into space from the poles. On the surface of the Sun there are great eruptions of plasma, which spurt out gusts of charged particles. If Earth happens to be in the path of these particles, the magnetic field will attract them and follow the currents to both the North and South Pole.

As billions of highly charged solar particles enter Earth's upper atmosphere, known as the thermosphere and mesosphere (between 80 and 640km high), they collide with atmospheric gases. This interaction excites these gases and they emit photons – small bursts of energy in the form of light. When there is enough of this photon energy, it is released and we start to see the Northern Lights dancing in the sky. The different colours we see depend on the type of gases present at the time, which can be influenced by how high up the aurora appears.

The most common colours of the aurora are green and red. These are created by oxygen molecules in our atmosphere. Oxygen at about 90–100km gives the vivid green and yellowish colours we see. Oxygen much higher up at about 320–350km into the atmosphere will give off the red colour as it is excited by the solar electrons. Nitrogen molecules present also get excited and the photons released will give off red and purple colours. Of course, you'll get a whole range of different shades of these greens, yellows, reds and purples depending on how much energy the aurora has and how the wavelengths of light blend with each other. While the naked eye can see the wonderful colours of the aurora, people who know about proper photography can use settings that will pick up the different wavelengths of colour much more vividly. This is why sometimes photographs of the aurora can look much more dramatic than if you were observing it for yourself.

Where Can You See It?

We've already established you can see the Aurora Australis around the Antarctic Ocean but places to visit on land are much more limited than those in the Northern Hemisphere. As a general rule of thumb, the closer you are to the Arctic Circle, the better. That's not to say you can't see it at lower latitudes into Europe and North America – it depends on the strength of any solar wind. The winter months of either hemisphere are the best months as this is when the dark nights are longer, giving you a much better chance of catching the aurora. In the Northern Hemisphere, aurora viewing season generally runs from October to May. You will also need to go somewhere that is very dark. Any light pollution from cities or towns will completely dampen the effect of seeing the aurora.

Can It Be Forecast?

Absolutely! The aurora is highly dependent on the Sun's eruptions spewing out the charged particles. Our understanding and monitoring of the Sun, and indeed space in general, has improved greatly over the last decade. Satellites are continually pointed at the Sun's surface so scientists can monitor solar activity. One of these satellites is called the

Advanced Composition Explorer (ACE) and is positioned around 1.5 million km away from Earth towards the Sun. It monitors the Sun's surface for Coronal Mass Ejections, solar flares and geomagnetic activity in real time. Once a Coronal Mass Ejection is spotted on the Sun it can take between one and three days for the solar wind to reach Earth. Space weather forecasters can calculate an expected aurora in that time frame. One prediction method space forecasters use to give an indication of the strength of geomagnetic activity reaching our atmosphere is the Kp (planetary) index. This index is monitored in real time using a number of observations. The index is on a scale from 0 (very quiet activity) to 9 (intense storm). The higher the number, the greater chance you'll see the aurora at lower latitudes. For example, looking at the diagram below, a storm with Kp-index of 5 would be strong enough to potentially see the aurora as far south as northern England and Wales. It's also worth noting that the higher the Kp-index, the more likely there'll be other consequences to our communications and power networks.

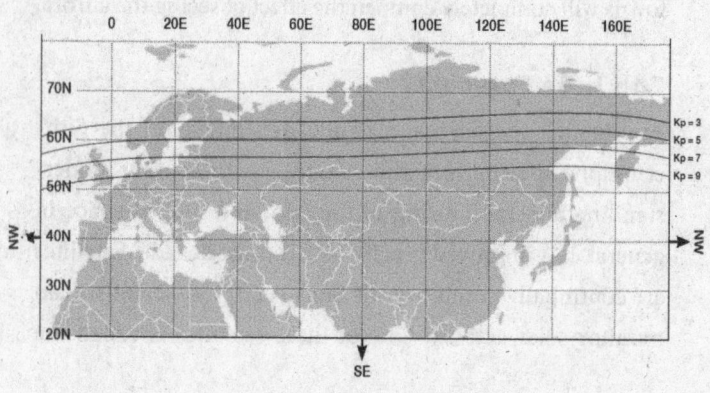

WATERSPOUTS

Waterspouts can look like tornadoes but over water. It's not quite as simple as that as while some waterspouts have tornadic features, the most common type of waterspout spotted is what we call a 'fair weather' waterspout. Either way, the horizon of water surface, thick cloud ahead and a menacing rotating column of air connecting the two looks very impressive. As waterspouts form over water, they don't tend to cause much damage and if you just happened to be on a boat out on the water, most are relatively small, so you'd be very unlucky to be getting in the way of one. On occasion, a waterspout can start its life over water but then travel on to land. If this were to happen, it might be obvious, but it would become a 'landspout' or 'tornado' depending on its initial formation.

Fair Weather Waterspout

The most common type of waterspout is a fair weather or non-tornadic waterspout, so-called as they don't form from a vigorous cumulonimbus cloud, unlike a tornado. These waterspouts are often quite long and thin, with wind speeds of less than 60mph and last typically no more than 15 minutes. This type of waterspout has very similar characteristics to that of a dust, fire or snow devil, as they originate at the surface before climbing skyward in a funnel of air that contains water, or more accurately, water vapour. Fair weather waterspouts tend to form mostly in the tropics and subtropics from August to October when sea/water temperatures are at their highest. There needs to be a lot of moisture in the atmosphere and most often developing cumulus clouds for them to form. If some cold air passes over the top of the warmer water, which will have a layer of warm air above it, it all becomes unstable and you start to get some updrafts and spinning from the local winds. A dark spot appears on the water surface and then with further development, a spray ring will occur. This is where sea spray will be lifted from the surface and rotate around the dark spot. If conditions continue to be right, a column of spinning air will rise further into the sky, eventually meeting the cloud overhead. The funnel appears hollow, encased by water vapour, and can look very dramatic as it snakes from the water to cloud. Given its appearance, many think the waterspout is comprised of seawater spinning around the vortex. However, it is not seawater, it's actually condensed atmospheric water, i.e. cloud. After a short time, the feed of warm water will disappear and the funnel and spray ring dissipates. While fair weather waterspouts are generally quite weak, some can grow quite large and could pose a danger to ships or aircraft unlucky enough to be operating in the area.

Tornadic Waterspout

As the name suggests, these waterspouts are bigger, more violent and are formed just like a normal tornado as a big cumulonimbus or supercell storm system generates the spin

Tornadic Waterspout

needed at the bottom of the cloud for a funnel to start. As in normal tornado generation, the funnel can extend down and grow larger with very strong winds. Eventually, the funnel cloud will touch the water surface as a tornadic waterspout with the distinctive spray ring. While you wouldn't want to get in the way of any type of waterspout, this type can be particularly damaging. They tend to be bigger with stronger winds, potentially up to 150mph, and could easily damage boats or anything else that happens to be on the water. The biggest concern with tornadic waterspouts is if they move on land. At this point, it would become a normal tornado with the potential to cause extensive damage to properties and people.

Raining Fish and Frogs

Some of the bigger waterspouts may be powerful enough to suck up fish and in some cases frogs from the sea or a lake. If this happens and the waterspout moves on to land, as it dissipates the fish can be rained out on to land. There have been a number of documented cases where this has confused communities with the sudden appearance of fish out of water on the streets. If the waterspout is particularly strong, it can transport frogs, fish of other wildlife right up the waterspout and into the cloud above. Here, they can get caught up in the updrafts of the cloud and remain there for much longer. As the cloud travels onshore, and in some cases many miles inland, gravity will eventually take over and the fish or frogs can be 'rained out'.

BROCKEN SPECTRE

Atmospheric phenomena and optics can often give us fascinating and beautiful weather effects, but sometimes they can bring something a bit spookier. The term Brocken spectre was first used in 1780 by a German natural scientist who often climbed the Harz Mountains in Germany. When he reached the top of the highest peak, the Brocken, he spotted a ghostly figure in the mist. It had a halo of colourful rings like a rainbow and towered in the distance. He called it the 'spectre of the Brocken'. Far from being a spectre, it is actually the shadow of the person observing it cast on to a cloud or mist at the top of a ridge, which often magnifies the shadow. For this to occur, the Sun has to be behind the person and, as well as giving a shadow, it creates a glow and colourful rings known as a 'glory' around it. The glory is essentially the reflection of the sunlight on the water droplets of the mist or the cloud. The exact physics of how a glory is formed are very complicated and we won't attempt to bamboozle you with the detail. All you need to know is that the sunlight (which you'll remember is white) from behind you hits the water droplets. It is then reflected and scattered within the droplet back towards your eye as distinct rings of blue, green, red and purple. You can also see a glory if you fly and sit in the window seat, looking down at the clouds. If you're on the opposite side of the Sun and there is cloud beneath you, the plane's shadow is cast on to the cloud with a glory around it.

THE ULTIMATE SUNRISE AND SUNSET

Stunning weather imagery is an important part of our weather presentation. In particularly, viewers photos that we showcase daily to tell the weather story. The BBC has a 'weather watchers' club, and they submit beautiful photos every day. The most popular shot is the sunrise or sunset vista glowing with deep pinks, reds and oranges

To understand what creates a good sunrise or sunset, let's discuss those vibrant colours. Light from the Sun contains the full spectrum of colours, which combine as white light. As the white light passes through our atmosphere, ice, water droplets and molecules of gas scatter the white light into the different colours, each with a slightly different wavelength. The blue colour has a short wavelength and is scattered more strongly than the colours at the longer wavelengths. This means the white light from the Sun is scattered all over the sky and it appears blue to the human eye.

During a sunrise or sunset, the light from the Sun has to travel through more of the atmosphere before it reaches our eyes. This means there are more molecules of gas, ice and water droplets to scatter the white light. The blue part of the light spectrum is scattered so much that we're just left with the large wavelengths of reds, oranges, yellows and pinks that reach our eyes, which is why the sunset and sunrise appear redder.

What Makes a Good Sunrise Great?

One big consideration is fairly obvious: the weather conditions. If there is a lot of cloud and/or precipitation, when you wouldn't normally see the Sun, you're not going to get a very good sunrise or sunset to view. Having uninterrupted easterly facing views (for a sunrise) or westerly views (for a sunset) is also important. Some of the best sunrises and sunsets are at the coast, where you can see well into the horizon for the moment the Sun either rises or sets below. The sea can also act as a reflector of those vibrant colours to enhance the whole picture. Equally, heading to high ground can also provide those uninterrupted views of the horizon.

While you don't want thick cloud blocking out the Sun in the first place, to have a great sunrise or sunset some cloud in the sky can help, particularly if it is high cirrus or altocumulus cloud. These types of cloud are made predominantly of ice crystals, which can reflect more of the reds and oranges beneath them and make it look like the whole sky is on fire.

Can We Predict a Good Sunrise or Sunset?

The saying 'Red sky at night, shepherd's delight. Red sky in morning, shepherd's warning' first appeared in the Bible and was said to give shepherds an indication of the expected weather conditions. There is some substance to this because in the UK, our weather normally comes in from the west. Therefore, if the Sun is rising in the east and shines on approaching cloud in the west, it'll give us vivid reddish colours and so warn us of an approaching weather system. If you have red sky at night, then the setting Sun in the west is shining on clouds moving away to the east and so fine, clear weather would be expected the next day.

To predict when there'll be either a good sunrise or sunset, we can use some of this saying. For example, if we know there's an approaching weather system, there might well be a good sunrise. We can also predict when there'll be a lot of the cirrus or altocumulus in the sky, which can enhance a great sunrise or sunset.

What About Dust and Pollution?

Pollution in the atmosphere can sometimes enhance the vibrancy of a sunrise or sunset. However, it can be dependant on the type of pollution in the atmosphere. If some of the particulates of pollution are large they can absorb more light and muddy the colours. You might not get the vividness of one or two colours, such as red or pink you might sometimes want in a good sunrise or sunset. In fact, this is also true during the day. If there is a lot of pollution, the sky appears to be a very hazy, whiteish rather than the vivid blue sky you'd expect when the air is really clean.

Dust acts in a very similar way as these are often quite large particles and don't scatter the colours like water vapour and ice crystals. But what dust can do is make the sky turn reddish orange in its own right. Back in October 2017, Hurricane Ophelia tracked to the west of the UK and Ireland, and as it did so, it dragged a large amount of dust from the Sahara and smoke from wildfires raging in Portugal and Spain. The more dense particles did the job of scattering out more of the blue light, leaving the reds and oranges to reach our eye. It was almost like a sunrise or sunset but during the middle of the day. The red Sun above the UK certainly caused a lot of interest, with tales of an apocalyptic sky.

In a digital age, we have noticed many more weather phenomena being captured and shared on social media. We have a fascination with the weird and extraordinary things spotted in the sky, whether it be strange clouds or an optical illusion. As meteorologists, we love it when people capture strange and rare phenomena without even really knowing it – allowing us to delve into explaining and sharing our own fascination with them. Curiosity around the sky and weather has been evident since ancient times when describing atmospheric phenomena with prophecies or being weather lores have been documented. Today, we have a greater understanding of the science behind the phenomena but we can all still be amazed by these weird and wonderful things.

WEATHER, SPACE AND PLANETARY INFLUENCES

WHAT'S THE WEATHER LIKE ON OTHER PLANETS?

The weather on each planet is essentially determined by a number of factors: shapes of orbit around the Sun, the distance from the Sun, length of day, the tilt of the planet (when there's more tilt, there's more variation in weather across the globe), and planetary atmosphere (or lack of one). For the faraway ice giants, Uranus and Neptune, their inner heat also drives circulation patterns beyond their surface.

Mercury

Mercury is testament to the fact that there are few environmental benefits in being the closest to the Sun. This nondescript greyish planet is the smallest in the Solar System and its thin atmosphere is incapable of capturing any heat. This means there are no clouds, rain or wind and the temperature range is extreme: from scorching days to bitter nights. Mercury's very slow spin means the Sun only rises every 176 Earth days. In stark contrast, it orbits the Sun with great speed, completing an orbit in 88 Earth days.

> **MERCURY DASHBOARD**
> **Distance from Sun:** 57.9 million km
> **Length of day:** 4,222.6 hours
> **Tilt:** 0 degrees
> **Mean temperature:** 167°C
> **Atmosphere:** very thin
> **Mass:** 0.06 Earths
> **Moons:** 0
> **Weather:** moonlike, no weather due to its very thin atmosphere

Venus

Venus may be further away from the Sun than Mercury, but it is the hottest planet in the Solar System. From a climate perspective, Venus is seen as having a runaway climate effect with a thick carbon dioxide atmosphere, plagued by acidic clouds that trap heat – this is global warming at the extreme. Violent upper winds rage around the planet, keeping the yellow and white clouds in motion and obscuring the surface. Venus has a very slow spin on its axis, the slowest out of all planets, and so the Sun rises just twice a year. Another oddity is that its sluggish rotation is actually in a clockwise direction, resulting in an almost spherical shaped planet. It has a shorter year than Earth and with a 3-degree axial tilt, variations in weather conditions across its entire surface are slight.

> **VENUS DASHBOARD**
> **Distance from Sun:** 108.1 million km
> **Length of day:** 2,802 hours
> **Tilt:** 3 degrees
> **Mean temperature:** 464°C
> **Atmosphere:** very thick
> **Mass:** 0.8 Earths
> **Moons:** 0
> **Weather:** very hot – global warming at the extreme

Earth

Earth is the only planet that we know of where life can not only survive but thrive. This is because of the life-sustaining mix of atmospheric gases that envelopes it: a delicate balance of nitrogen, oxygen, hydrogen, carbon dioxide, water vapour and other trace gases. A thin shield of ozone above the weather-making layer of the atmosphere protects us from harmful UV rays. Below, the Sun's light transforms into heat as it hits Earth's surface, allowing air to rise and fall. These zones of high and low air pressure, forever in motion, produce corresponding dry and wet weather. The redistribution of this heat and moisture across thousands of miles through a network of wind and ocean patterns moderates extreme environments, further enabling a diverse set of ecosystems. Earth's 23.4-degree tilt allows for seasonal differences through the year, adding to the broad and diverse set of weather conditions worldwide.

> **EARTH DASHBOARD**
> **Distance from Sun:** 149.6 million km
> **Length of day:** 24 hours
> **Tilt:** 23.4 degrees
> **Mean temperature:** 15°C
> **Atmosphere:** thin
> **Mass:** 5.97 x 10-24kg or 1 Earths
> **Moons:** 1
> **Weather:** great balance for human life

Mars

Mars, also known as the Red Planet, can be described as a cold desert. In some ways it is similar to Earth, with both planets having ice caps, seasonal variability (due to a 25-degree tilt) and significant weather patterns. Mars is further from the Sun and has a much thinner atmosphere and its climate has changed considerably over billions of years, with some scientists suggesting it was once able to sustain water. The planet now exhibits an extreme in climate change, with an abundance of carbon dioxide and a dry environment that cannot hold much water. It has a north and south pole consisting of ice and frozen carbon dioxide. The iron content on its surface gives Mars its reddish colour and its extreme elliptical orbit and axial tilt produces huge variations in conditions, which often manifests in violent dust storms. The atmosphere of Mars may be thin, but it is laden with clouds, wind and dust. Temperatures are a lot lower than Earth, but across the planet there is an extreme temperature range; at the poles temperatures can dip as low as -125°C but as high as 20°C during the daytime close to the equator.

> **MARS DASHBOARD**
> **Distance from Sun:** 227.9 million km
> **Length of day:** 24.7 hours
> **Tilt:** 25.2 degrees
> **Mean temperature:** -20°C
> **Atmosphere:** very thin
> **Mass:** 0.107 Earths
> **Moons:** 2
> **Weather:** dust storms

Jupiter

One of two gas giants, Jupiter hits the top spot in terms of size – 11 times larger than Earth. It is also the fastest-spinning planet, which means flatter poles and a bulge at its equator. The faint Jovian ring system is made up of three bands of dust, but these are far less visible than the raging clouds and storms that give this planet its distinctive look. Jupiter has a gaseous atmosphere that sits above an ill-defined but denser interior. The atmosphere is mainly hydrogen with a smaller percentage of helium and other elements. Jupiter has a huge range in temperature, with an average of around -145°C just below the cloud layers but a core estimated to be around 24,000°C. This is key to heating the rest of the planet through the process of convection (rising heat). Although Jupiter has no distinct seasons, with just a 3-degree tilt, it does display violent storms. The skies are composed of thick ammonia crystal clouds. These are banded latitudinally into layers of yellow, brown and white; the darker belts are where the atmospheric motion is descending and the lighter zones where the air is rising. Where these different bands interact, storms develop. The most notorious is the Great Red Spot – a hurricane with winds in excess of 400km/h (260mph). The size of at least two Earth planets, this Great Red Spot has been present for over 400 years, although there is evidence that it is shrinking. Other weather phenomena include lightning displays and auroras, so it's fair to say similar weather types to Earth, but, as with everything associated with Jupiter, the elements are a lot grander and far more powerful.

JUPITER DASHBOARD

Distance from Sun: 778.6 million km
Length of day: 9.9 hours
Tilt: 3.1 degrees
Mean temperature: -145°C
Atmosphere: very thick
Mass: 318 Earths
Moons: 67
Weather: stormy

Saturn

It may not be as large as Jupiter, but Saturn is still categorised as a gas giant. It is further still from the Sun with an average temperature of -178°C. There are four planets that have a ring system: Jupiter, Saturn, Uranus and Neptune, however Saturn's is the most visible. Four sets of icy rings adorn this planet, and with over 62 moons, the Saturn system is a fascinating study for scientists. Saturn's weather is equally intriguing. Its extreme 27-degree tilt results in seasons lasting over seven years, although it has the second shortest day in the Solar System, being a mere 10.7 hours long. Saturn has thick layers of yellow, grey and brown clouds containing ammonia (upper layers) and ice (lower layer). These layers are driven around the planet by intense jet streams and violent storms. Like Jupiter, it has its own huge storm spots, where the winds can reach 1,000mph. It also has many lightning strikes. The Voyager mission in 1981 revealed a hexagonal-shaped cloud pattern spanning 12,700km that crowns Saturn's north pole – a structure not found on any other planet. Currently, there are more questions than answers as to why this fascinating geometric body of fluid and wind exists.

> **SATURN DASHBOARD**
> **Distance from Sun:** 1,433.5 million km
> **Length of day:** 10.7 hours
> **Tilt:** -26.3 degrees
> **Mean temperature:** -178°C

> **Atmosphere:** very thick
> **Mass:** 95 Earths
> **Moons:** 62
> **Weather:** storms and lightning, winds up to 1,000mph

Uranus

The first of two ice giants, Uranus appears blue/green due to the methane content in its atmosphere. It also has a set of faint rings; to date, there are 13, comprising of dust and larger particles. This ice giant has the lowest-recorded temperatures of any planet. Temperatures can dip as low as -224°C. However, with a greater range of temperature the overall average generally remains higher than the most remote planet in our solar system – Neptune. Scientists suggest the very low temperatures are due to Uranus' odd orientation: it spins on its side with an extreme tilt of 98 degrees and takes 84 years to revolve around the sun. This also means Uranus has decades of permanent daylight during its long summer season and winter nighttime, allowing for this greater range in temperature over a long period of time. Uranus has a weaker internal heating system than Jupiter and Saturn, so its atmospheric dynamics are distinctly less active than other planets. But, weather systems do rotate in bands of icy methane clouds carried across its atmosphere by strong winds. Occasional dark storm spots appear on the planet as well, and there are still a lot of questions as to why they form.

> **URANUS DASHBOARD**
> **Distance from Sun:** 2,872.5 million km
> **Length of day:** 16.1 hours
> **Tilt:** 97.8 degrees
> **Mean temperature:** -195°C
> **Atmosphere:** thick
> **Mass:** 14.5 Earths
> **Moons:** 27
> **Weather:** extremely cold with strong winds

Neptune

This ice giant is furthest from the Sun and has, by far, the strangest weather of all planets in the Solar System. Most of its dense mass is a combination of methane, ammonia and hydrogen. It's the presence of methane that absorbs red light and leaves blue light which creates Neptune's distinct blue colour. The appearance of white dots and dashes across Neptune's surface are from high cloud composed of frozen methane. It has at least five faint rings; these are made of dust and rock, making for spectacular imagery – although this is the only planet that cannot be seen by the naked eye from Earth. Neptune's axis of rotation is tilted at 28 degrees, similar to Earth, and like our blue planet, this blue ice giant has seasons, or periodic variations in its weather patterns. These seasons though can last over 40 years, as it takes 165 Earth years to orbit the Sun. The Sun drives Earth's weather and wind patterns – Neptune receives 900 times less sunlight yet it has intense storms, by its dark spots, and extreme winds dominate the weather. In fact, this remote planet is the windiest of all eight, with wind speeds of over 2,000km/hr or 1,243mph. Scientific debate still throws up many questions about Neptune's weather, with some scientists suggesting that it is not only influenced by the Sun, but perhaps more so by galactic cosmic rays or high-energy particles from outer space.

> **NEPTUNE DASHBOARD**
> **Distance from Sun:** 4,495.1 million km
> **Length of day:** 16.1 hours
> **Tilt:** 28 degrees
> **Mean temperature:** -200°C
> **Atmosphere:** thick
> **Mass:** 17.1 Earths
> **Moons:** 14
> **Weather:** cold, dark and windy

HOW DOES EARTH'S ATMOSPHERE SHIELD LIFE?

The atmosphere acts as a shield to sustain life. There are a key combination of gases essential to the protection and survival of the planet's biospheres and ecosystems. Approximately 78% of the atmosphere is nitrogen, 21% is oxygen and the remaining 1% is made up of water vapour, carbon dioxide and other trace gases.

How Do the Individual Layers of the Atmosphere Protect Earth?

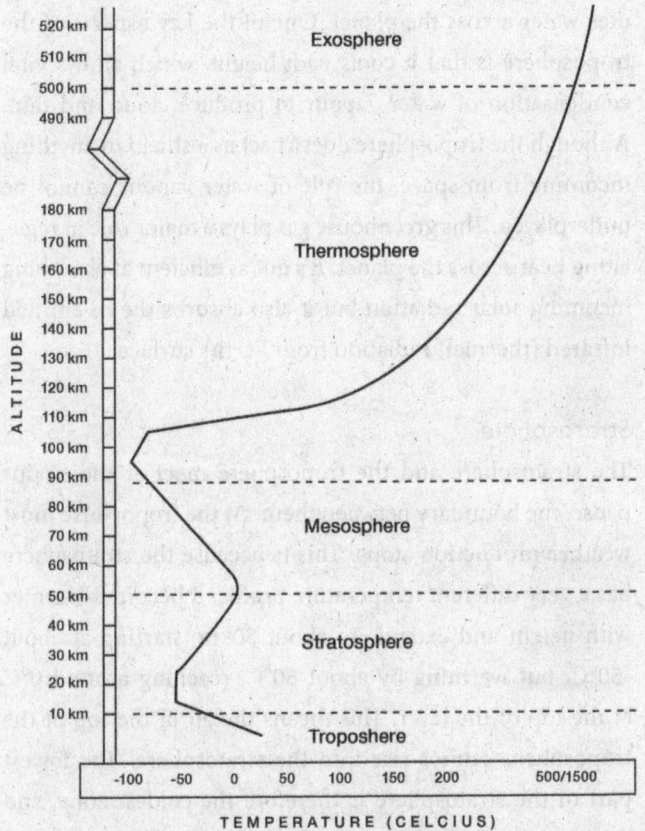

Temperature profile of Earth's atmosphere with height

Troposphere

The troposphere is the weather-making layer. At its deepest, it is about 13 miles or 20km deep and touches Earth's surface.

This critical layer sustains life by circulating heat and water vapour that in turn moderates the temperature and distributes water across the planet. One of the key aspects of the troposphere is that it cools with height, which allows vital condensation of water vapour to produce cloud and rain. Although the troposphere doesn't act as a shield to anything incoming from space, the role of water vapour cannot be underplayed. This greenhouse gas plays a major role in regulating heat across the planet. It's not as efficient at absorbing incoming solar radiation but it also absorbs the re-emitted infrared (thermal) radiation from Earth's surface.

Stratosphere
The stratosphere and the troposphere meet at the tropopause, the boundary between them. At the tropopause most weather production stops. This is because the stratosphere has a very different temperature profile; it becomes warmer with height and extends to about 50km, starting at about -50°C, but warming by about 50°C, reaching around 0°C at the top of the layer. This means the air at the top of the troposphere cannot rise into the stratosphere. The lowest part of the stratosphere is therefore the coldest zone, and this is where the Stratospheric Polar Vortex (SPV) sits, even though it is the mechanics of the troposphere that result in the distribution of heat and moisture around the globe. The SPV plays its own part. It develops over the poles during the dark winter, continually cooling down to the lowest point through the end of December. The presence of the

bitter vortex that drives strong winds around the poles maintains a temperature gradient with the low latitudes and fuels a strong mid-latitude jet stream. It's this jet stream that transports the excess heat across the equator northwards, moderating temperatures here and in doing so, transferring energy and moisture across the globe. The more famous, and far more effective, part of the stratosphere that creates a shield around Earth is the ozone layer. It only contains less than ten parts per million of ozone or O_3, but this is more concentrated than other parts of the atmosphere. It's here that 97 to 99% of the incoming harmful UV radiation is absorbed and the absorption of solar radiation allows this layer to warm up.

Mesosphere

The mesosphere cools down rapidly to -100°C. This is because the rate of absorption of solar radiation falls. The atmosphere here is incredibly thin and there is no ozone. It's a thick layer that is 35km deep in places. The mesosphere plays host to spectacular displays of shooting stars. These are meteors that hit this layer and burn up before hitting Earth. As such, the mesosphere provides a protective layer against most rocks and other debris that fall from space and get sucked into Earth's gravitational field. Meteors make it through the exosphere and thermosphere because there is not much in these layers to slow them down, but thankfully, the mesosphere has enough gases to slow down the progression. Friction from this slowing down creates heat and most rocks normally break down and vaporise. It's not the case every time, as the pot-holed landscape of many continents can testify. The mesosphere is responsible for millions of lives saved, possibly on a yearly basis.

Thermosphere

Next to the coldest layer lies the hottest layer. Temperatures here can reach 2,000°C. This is due to the absorption of ultraviolet and X-ray radiation from the Sun. Interestingly, its remote position 500 to 1,000km from Earth's surface and gas composition means it is the home to stunning auroras or what is known as the Northern/Southern Lights, where the Sun's charged particles collide with atoms of oxygen and nitrogen to produce an incredible coloured light display.

Exosphere

The exosphere extends from the boundaries of the thermosphere up to 10,000km, becoming the gateway to outer space and the home to many man-made satellites. The thermosphere's extremely high temperature profile influences this exterior shell of the atmosphere, resulting in a huge temperature range from 2,000°C to 0°C, and the air here is coldest at night. The atmosphere also is very thin yet still traces of gases, such as oxygen and carbon dioxide, can be found here. This is the final zone of the atmosphere that is influenced by Earth's gravity. Importantly, it is the planet's first defence against solar radiation and the first to connect with any object flying earthbound.

ARE ASTEROIDS EARTH'S BIGGEST THREAT?

Stephen Hawking, in his final book published in 2018, *Brief Answers to the Big Questions*, discusses his thoughts on one of the biggest threats to Earth – an asteroid collision. It's something we have no defence against and possibly the reason why dinosaurs were wiped out 66 million years ago. Even though Earth has an impressive defence system, with layers of atmosphere all primed with specific purposes, its potted landscape of craters is testament that flying rocks in space do penetrate and some impact the surface. Satellite observations suggest that between 5 and 300 metric tonnes of cosmic dust enter the atmosphere each day, and occasionally this is picked up in footage of burning debris flying through the sky. But let's first debunk the terms:

- *Asteroid:* a large rocky body in space without any atmosphere but in orbit around the Sun. This body is defined as greater than a metre wide and at times up to hundreds of kilometres across. Ceres is the largest observed asteroid, with a diameter of 940km, so large it's now considered a dwarf planet. 4 Vesta comes in second at 525km in size. Between Mars and Jupiter lies the most populated part of the Solar System, where the main belt of asteroids orbit the Sun. It has been widely suggested that Near Earth Asteroids are ones that were originally deflected off course from this asteroid belt and that there

- *Meteoroid*: any rock in space that is less than 1m wide is known as a meteoroid. These are much smaller rocks or particles in orbit around the Sun. They do collide with Earth's atmosphere regularly. If a meteoroid enters Earth's atmosphere and vaporises, it becomes a meteor, which is often called a shooting star. Earth's orbit coincides on a frequent basis with debris from the tail of a comet, and incoming meteors bring a dazzling display of shooting stars. The consistent orbit of such comets, such as the Swift-Tuttle, means that these shooting star events such as the August Perseids meteor shower can be well forecasted every year.
- *Meteorite*: if larger meteoroids or asteroids survive the fiery passage through Earth's atmosphere they are then called meteorites. Thousands of meteorites don't make impact on the surface of Earth, burning up in the lower atmosphere and momentarily lighting up the sky. Occasionally, pieces of space rock, less than half a kilogram, do survive the journey, hit Earth and go virtually unnoticed. The most significant observation that many have witnessed is the aftershock of energy as a large meteorite hurls to the ground. This effect can shatter windows, stir up bodies of water and disturb forests. However, Earth has been dented by numerous larger meteorites that create huge craters on the landscape. These are incredibly rare events spanning tens of thousands of years.

- *Bolide*: a very bright meteor that often explodes in the atmosphere, sometimes termed a fireball. These can create shockwaves as they enter the lower atmosphere and make impact. In February 2013, a bolide, estimated to be 18m wide, exploded through the sky, shattering windows over Chelyabinsk, Russia, and injuring 1,200 people.

Evidence of Past Impacts

The largest observed crater formed by an asteroid is the Vredefort crater in Free State, South Africa. The impact date is estimated to be 2 billion years ago and the crater has a radius of 190km. Geologists studied in great detail the concentric semi-circular ridges that were still engraved into the sides of the crater, revealing tell-tale signs of impact. It is thought that the asteroid was between 5 and 10km in diameter; just these dimensions alone display the immense energy that an asteroid can still maintain on impact. No other crater rivals this, but there are many across the globe that are still very impressive.

The Chicxulub crater in Yucatán, Mexico, measures 180 to 240km in diameter. On a geological timeline this particular crater is relatively young, estimated to have been formed from an impact 65 million years ago. There is strong reasoning that this particular asteroid was responsible for the extinction of the dinosaurs. The might of this asteroid collision sent shockwaves through the biomes of Earth, propelling volumes of dust and particulates through the atmosphere covering the whole planet and sparking gargantuan tsunamis, enough to engulf neighbouring land masses. Yet it remained undiscovered until the late 1970s when geological surveyors from a Mexican oil company came across the site. Over millennia the Chicxulub crater's original depth of 900m has filled in and now dips a matter of metres. More recently, in December 2018, a fireball was observed diving into the Bering Sea. It was picked up on satellite imagery, shining as 'bright as Venus', masking the intense energy it produced – a huge 173 kilotonnes and 13 times larger than the Hiroshima bomb of 1945 – as it detonated 26km above Earth.

What Would Happen If a Large Asteroid Hit Earth?

Calculations indicate that asteroids greater than 25m but less than 1km would cause damage on a local scale. Anything larger than 1 or 2km would have consequences on a global level. The most lethal repercussions from a large asteroid impact would be wind blasts and shock waves. A spike in air pressure could rupture internal organs and wind blast would hurl bodies and flatten buildings and forests. Other damaging effects would be intense heat, flying debris, tsunamis, seismic shaking and annihilation from the direct impact and cratering. Like other large objects in space, asteroids are influenced by gravitational forces, so they have their own orbit. This makes their path relatively predictable. Cataloguing Near Earth Objects (NEOs) is a titanic task; space is very crowded, and it seems to be getting even more crowded with every decade. Mapping NEOs against a background of other debris orbiting in space could be described as finding a needle in a haystack, but astrophysicists have made great strides to do just that.

How Often Do Asteroids Get Through Earth's Atmospheric Shield?

Every year or so, an asteroid the size of a car smashes into Earth's atmosphere, igniting due to friction and creating a bright fireball before burning up prior to hitting land or sea. Sometimes they do make impact. Every 2,000 years or so, a meteorite the size of a football field hits Earth and causes significant damage to the area. It's estimated every few million years, an asteroid large enough to devastate a region of Earth or impact on a global scale occurs. There are a growing number of experts who are now suggesting that these estimates are conservative. In part, this is because Earth is more water than terra firma so it does make sense that there could have been more significant impacts that dived to the depths of the ocean. There may be a number of craters hidden beneath the ocean surface that have eroded and morphed with time. Also, smaller craters and holes could easily have weathered, eroded or been covered up by volcanic eruptions through millennia. This is one reason why the Moon looks far more potholed than Earth – the marks of collisions remain for millions of years. In fact, Earth attracts more meteoroids and asteroids than the Moon due to its stronger gravitational force (it has a bigger body and therefore a larger mass).

HOW DOES GRAVITATIONAL FORCE AFFECT EARTH?

The force of gravity is a prerequisite for life. Every object is attracted to another by way of gravitational force. Without this force there would be no structure to space, nothing would exist as we know it in the universe today. The mass of an object determines how strong that force is, so the Sun, as the largest body in our Solar System, exerts the greatest gravitational force, attracting all other planetary bodies.

Moon Phases

The gravitational force also depends on distance – the further two bodies are from each other, the less this force is. That is why the Moon plays an important part in Earth's tides. The combination of the Sun and the Moon pulls the body of water in and out, depending on their position to each other. A full Moon and new Moon tells the observer that the Sun and Moon are aligned; the full Moon being completely exposed to the light of the Sun, and the new Moon occurring when Earth casts a full shadow over the Moon. Both are due to the alignment of the Sun and Moon exerting the largest gravitational pull on Earth. When a portion of the Moon is illuminated, this is because the Moon is at an angle to the Sun and there is a weaker force. When they are aligned, this force accumulates and is strong and so results in higher spring tides. When they are not aligned, the two forces oppose each other and the resultant force is far weaker, leading to lower neap tides. The highest tides of the year are during the Equinox when the sun is directly over the equator, and when the Moon and Earth are aligned, so spring tides at Equinox produce the greatest tidal range. The opposite can be said for the solstice coinciding with the neap tides and leading to a minimal tidal range.

How Does Gravity Affect the Air?

Gravity is necessary to hold the atmosphere in place. Without it, the air would escape into space. Gravity acts downwards and it's this downward force that is an intrinsic part of every atmospheric process. At Earth's surface air is compressed by the weight of air above it and this causes the air closest to the ground to exist at a greater pressure than at higher elevations. The air becomes thinner with height to a point where humans have no ability to breathe.

There are local effects that challenge gravity. When the Sun is strong and the ground warms quickly, the air close to the ground warms up and becomes less dense and therefore rises. The air is rapidly replaced by surface level air from the surrounding area; which is then again warmed and rises. This environment will produce its own weather, sometimes cloud and showers, an increase in wind and a change of temperature. It is a localised area where the pressure is lower, or in another way, a zone where the air rises. As the Sun sets and the air cools so pressure builds at the surface as the air loses energy and descends. The change in air pressure is a key factor in determining local conditions, and like other atmospheric processes, gravity is always present and yet also often challenged.

THE TECHNOLOGY OF WEATHER

SHOULD YOU ONLY RELY ON AN APP OR WEBSITE FOR THE WEATHER FORECAST?

In an increasingly digital age, we all rely more and more on our smartphones and speaking devices at home to give us the information we need. It has become much easier to access a weather forecast than it was in the 1990s when we had to either wait until after the news to watch the forecast, listen to it on the radio or read it in the newspaper. Twenty-four-hour news will now give you regular weather forecasts throughout the day and when we're on the move weather apps on our smartphones can give us a forecast for any town or city around the world.

Still, one common complaint we get is why the weather seemingly changes from hour to hour, day to day on the app or via a website? To fully answer this question, you need to understand how a weather forecast is created. Before we can forecast the weather in the future, we need to have a good understanding of what the weather is doing right now around the world. We can do this by taking weather observations at the ground of the temperature, wind, pressure, humidity and cloud cover. Weather balloons are released into the atmosphere and gather data from the surface to the top of the atmosphere.

Aircraft, ocean buoys, ships and satellites also gather huge amounts of meteorological data. In fact, every day, millions of bits of weather data are gathered from around the world and sent to world meteorological forecast centres such as the Met Office, European Centre for Medium Range Weather Forecasts, Japan Meteorological Agency and the National Weather Service in the US, to name a few. To turn this observational data into a forecast, scientists have developed mathematical equations and computer codes that are used to define how air behaves and shapes our weather around the world. This is called Numerical Weather Prediction and it uses complex physics and maths requiring a huge amount of computing power. Each of the major weather centres around the world has their own 'supercomputers' that are able to process the millions of observational data into forecasts from quadrillions (thousands of billions) of calculations every second. The result is a weather forecast from hours to days ahead.

There are also different types of weather models meteorologists use that can either give us one forecast solution known as deterministic or one that can give us 50 different forecast solutions. Ensembles are based on making small changes at the start and seeing how these would influence the forecast in days ahead. This is a useful tool to give us an idea of the certainty or probabilities in a forecast. If all 50 forecasts are similar, there is good agreement and we can be more certain of a weather forecast, but if all 50 forecasts are different then we can say we don't have a great deal of confidence in the forecast.

THE TECHNOLOGY OF WEATHER

Weather forecasting is complicated and it is far from an exact science, but with increasing technology and knowledge over the last three decades a four-day forecast in the UK is as accurate as a one-day forecast was in the 1980s. When it comes to weather apps, there are lots of different types all getting their data from different sources, each with their own algorithms on how to display the data. Some apps will show hourly data up to five days ahead, some will give you a daily forecast for 20 days ahead. Some only show a few weather elements, others will show a whole range of meteorological data. Some will show you the chance or probability of precipitation and others won't. Some weather apps have a human forecaster quality controlling the data, others use data straight from the computer models.

Different weather apps may show a slightly different forecast even for 12 hours ahead and may seem to change every hour. This is because algorithms used in these apps will also take observational data every hour to try and 'align the forecast', which is the same as looking out of the window and noticing that it is actually sunnier/wetter/drier than what the computer forecast model was suggesting. This could then have an influence on what the forecast is over the next six to 12 hours, with algorithms updated and potentially changing the weather symbols on your app. In the UK, the major forecast models are run every six hours so you may notice bigger changes in the forecasts then, especially when looking at the period of two to five days ahead when small changes at the start of the forecast model may result in greater changes towards the end of it.

CAN WE CONTROL THE RAIN AND WHAT IS CLOUD SEEDING?

For centuries, humans have been attempting to control the weather in some sort of fashion, whether trying to stop a hurricane smashing into a community, dispersing fog at allied airfields during World War II, making more snow at ski resorts or creating rain to help with drought and agriculture. Manipulating the weather isn't easy and so far the success and economic value hasn't been particularly good. However, making rain is seen as one type of weather modification that is worth the investment.

Some of the earliest documented cases of trying to make it rain are from the USA, where in the 19th century, businessmen were claiming they could make rain. Desperate farmers whose crops were struggling with drought paid money to use these rain-making services. Most of the time, there was no reputable science involved and these charlatans would take the money and head off to the next town of unsuspecting farmers. They soon gave up, but there has been an interest in manipulating the skies to produce rain ever since.

The earliest robust scientific experiments surrounding rain-making took place in 1946 when an American scientist called Dr Vincent Schaefer tried to make artificial clouds in a 'cloud chamber'. He added dry ice into one of his cloud-making experiments and found that the water vapour increased in the chamber and formed a cloud. The ice provided a nucleus, a seed, to which water droplets could attach themselves. In nature, it is ice, dust, salt or sand in the atmosphere that behaves as these seeds. Meteorologists call these cloud condensation nuclei (CCN). The theory is that if you can increase the CCN within a cloud, you can promote the growth of more water droplets and therefore there is more water available to be rained out.

Cloud-seeding methods soon came to the attention of governments and specifically, the military. They started to conduct their own research into ways of trying to make rain for military gain. In the UK, during the early 1950s, the Ministry of Defence started taking part in weather modification experiments. In declassified minutes from an air ministry meeting in November 1953, uncovered by a BBC Radio 4 history investigation, it was shown that the military were interested in increasing rain and snow by artificial means. Uses for this modification included 'bogging down enemy movement', and 'incrementing the water flow in rivers and streams to hinder or stop enemy crossings'. The experiments conducted by the Royal Air Force were known as Operation Cumulus and essentially involved releasing dry ice into clouds.

On 15th August 1952, Lynmouth in Devon was subjected to one of the worst flash flood events of the time. Thirty-five people died as water, mud and rock slides swept through the village. At the time, the general public thought this was just a horrific natural disaster and this belief wasn't questioned for years until declassified documents from the Ministry of Defence and the investigation by the BBC showed that Operation Cumulus experiments were taking place that week around the area. Documents also showed that as a result of the disaster Operation Cumulus was stopped very soon after. While there is still some controversy over whether the cloud seeding enhanced the rainfall, the BBC investigation revealed documents suggesting that the Air Ministry and the Treasury were very anxious and aware about the potential for rain-making to cause damage, not just to military targets and personnel, but to civilians too.

Why Do We Want to Cloud Seed?

If all this experimentation was stopped so swiftly, we ask why this discovery was so important. As well as the military advantages mentioned above, agriculture suffers whenever there is a lack of rainfall. Livelihoods would be destroyed if a farmer's crops didn't get enough water and failed. This is why USA farmers were so keen in the 19th century to pay businessmen to increase the rainfall. Even today, agriculture suffers when there is a prolonged drought. Increasingly though we now populate desert areas throughout the world, such as the United Arab Emirates (UAE), where there simply isn't enough water to service an increasing population. Technology surrounding cloud seeding has improved over the last decade and has become a big business. While it is difficult to accurately tell how much extra rain a cloud will give after being seeded, estimates put it at around a 15–35% increase in rainfall. More recently, while it has been effective in creating more rain in a particular area, there have been some instances where cloud seeding has been used to make a particular area dry for a period. One such example was in 2008, when during the opening ceremony of the Beijing Olympics organisers were so keen for it to be dry for the showcase event that the Weather Modification Office of the China Meteorological Administration were tasked to cloud seed an area downwind of the Bird's Nest stadium. The 'weather changers' fired 21 rockets filled with the cloud-seeding product silver iodide into the sky above the city. This triggered rain showers in Baoding city, southwest of Beijing, before they could reach the capital. On the night of the opening ceremony, it was dry.

The most common way of cloud seeding is to release the product directly into clouds via airplanes. Once scientists have identified the right type of cloud, which are known as cumulus shower clouds, on a day they want to enhance rainfall planes will fly straight into the cloud and use flares to release the cloud seed.

How Common is Cloud Seeding Globally?

We've touched already on the Chinese, who have the largest Weather Modification Office in the world. As well as the work they did during the 2008 Olympic Games, some of the ways in which they use weather modification include to prevent hail storms damaging crops and to make rain to counteract the effect of dust storms. Their biggest project to date is to try and alleviate the year-on-year drought of the Tibetan Plateau. This undertaking involves tens of thousands of 'burn chambers' that burn solid fuel to produce silver iodide. This then gets transported into the atmosphere with the hope of seeding the clouds and enhancing the rainfall. They aim to increase rainfall over an area three times the size of Spain, with up to 10 billion cubic metres of water a year.

There are over 50 countries who are known to have some kind of involvement in cloud seeding, with China, the USA and the UAE being some of the bigger players. Water scarcity is a big issue in the Gulf states, where they lack the water they need to service their growing populations and respond to climate change. The UAE currently uses cloud seeding when the conditions are right to produce more rainfall. Estimates from the National Center of Meteorology in the UAE suggest for every cloud-seeding operation costing $5,000, they can generate an extra 30% of water worth $300,000.

The UAE appreciates that the efficiency of cloud seeding is still relatively low, so it is also investing a lot more money in research around improving the effectiveness of its operations, making it a global leader in the future of technology in the field. Research is being undertaken to a) improve the cloud-seeding product and b) come up with completely new ideas to enhance rainfall. The use of nanotechnology is leading the way in improving the cloud seed, with results from chemical engineer Dr Linda Zou already suggesting a 300% increase in larger water droplets in a cloud.

Is Meddling with the Rain Ethical?

It has been accepted by the World Meteorological Organization that weather modification techniques are needed in some areas of the world where there is a real benefit in alleviating drought. In such circumstances, small-scale rain enhancement is seen as an understandable practice. However, there is a greater caution among the scientific and political communities surrounding much larger weather modification techniques, such as the Tibetan Plateau project. The atmosphere has no political boundaries and some argue that if you cloud seed on such a large scale and create more rain in one country, are you then 'drying out' the clouds before they may naturally rain out on to another country? It's an interesting question and debate will surely continue, while at the same time weather modification techniques may become more standard practice.

Is Cloud Seeding the Answer to Global Drought?

Organisations can clearly see the benefits of cloud seeding on a local scale to improve the amount of water available. In countries where they use cloud seeding to improve water supplies for agriculture and public needs, such as the USA, China and the UAE, there are clearly benefits. These are generally environments where there is crucially some sort of rainfall pattern already in their climate. For example, they have the clouds in the first instance to carry out cloud seeding to *enhance* the rainfall. A proper drought is defined as lack of rainfall over a long period of time, though. This means that in countries or areas that are officially in drought, they simply don't have enough moisture in the atmosphere to create clouds in the first place. The technology to create clouds doesn't currently exist. Therefore, while cloud seeding can help on a local scale, it is not the answer to larger areas and eradicating drought.

CAN WE CHANGE THE TRACK OF A HURRICANE?

While there have been instances of people modifying to seek military advantage, most weather modification practices today are in place to locally improve conditions. After all, the idea of trying to protect life and property from some of the most destructive weather forces doesn't seem too crazy, but the question is what is really possible? Could we actually divert or break up tropical cyclones before they make landfall, mitigating the need for mass evacuations, death, destruction and huge economic losses?

Stopping or diverting a hurricane in its tracks is no mean feat. The amount of energy contained within a typical hurricane is phenomenal. If you look at just the energy produced by the wind, it equals about half of the world's electricity production in a year. The energy it releases as it forms clouds is 200 times the world's annual electricity use and the heat energy within a hurricane is equivalent to a 10-megaton nuclear bomb exploding every 20 minutes. That's nearly 700 times more powerful than the nuclear bomb dropped on Hiroshima in 1945. Now, think about scaling up that energy to reflect the power of the bigger Category 4 or 5 hurricanes like Katrina and Irma.

Early Attempts

Attempts in hurricane modification go back to the 1940s when scientists in the USA started looking into how ice crystals could weaken storms. They called it Project Cirrus and in 1947, they flew a plane off the coast of Georgia to inject a hurricane with dry ice. The physics behind this is the same as it is in enhancing rainfall within a cloud: by initially encouraging more snow and rain to develop in the core of a hurricane you, in theory, take some of the energy away and essentially weaken it. While the hurricane in their experiment changed its path, it had nothing to do with Project Cirrus and no results could be proved so the project was canned.

After a number of devastating hurricanes hit the USA in the late 1940s and 50s, the United States government were keen to try hurricane modification again, so President Eisenhower commissioned an investigation into storm modification. In 1962, Project Stormfury was set up, which consisted of scientists at the National Hurricane Research Lab and the Navy Weapons Centre. The idea was very similar to Project Cirrus, but the Stormfury team had more advanced cloud-seeding technology and a better method of distributing it within a hurricane. Their plan involved releasing silver iodide particles in large quantities around the eyewall of the hurricane. After testing and further research, the team seeded Hurricane Debby in 1969 and found the sustained winds weakened by 15–30%. The problem was that it was very difficult to ascertain whether the winds weakened as a direct result of the cloud-seeding intervention or whether the hurricane would have naturally weakened because of the environmental conditions. Project Stormfury was cancelled in 1980.

While these early attempts at storm modification were inconclusive and eventually abandoned, to this day, after every major hurricane threatening land, people ask whether there is something we can do to stop these potentially devastating events. Some suggestions include dropping bombs within a hurricane. The problem is that because a hurricane has the same energy as many atomic bombs, unless you were going to nuke a hurricane, a conventional bomb wouldn't do much. And no, nuking a hurricane isn't a good option – you'd just create a nuclear hurricane with radioactive rain that would kill ecosystems and us!

So, as seeding and bombing hurricanes don't appear to be good options, the best way might be to manipulate the hurricane before it amasses too much energy. To do this we'd need to look at the areas where hurricanes form and how we could manipulate temperature, moisture and wind. In response to this, distinguished Scottish engineer Professor Stephen Salter has suggested that one solution could be to cool the ocean in the areas where hurricanes really start to intensify. His theory is that if you reduced the sea surface temperature below the critical 26°C, the amount of heat energy would also reduce, and consequently, the hurricanes would be weaker. On this basis, he designed the 'Salter Sink', a wave-powered pump to move warm surface water down to lower depths. However, it is thought that you'd need hundreds, if not thousands, of these pumps to cool a sufficient area of the ocean where a hurricane might cross. So, while the science behind this is sound, there are many questions about the feasibility of such a project.

Professor Salter also came up with the idea of making the clouds a bit brighter around the development of a hurricane. This is a similar principle to the seeding schemes of projects Cirrus and Stormfury, but the idea is to spray aerosols into the sky using hundreds of unmanned boats that could move into the paths of hurricanes. This would result in making the clouds brighter and increasing their reflectivity. In turn, this would reduce the amount of sunlight reaching the sea surface and therefore reduce the temperature.

Hurricanes can cause millions, sometimes billions, of dollars of damage when they make landfall, so it is not a crazy idea to invest money in trying new technologies to change the track or intensity of them. Nevertheless, it feels like we are still a long way off being able to control these powerful weather systems and maybe, for now, money may be better invested in forecasting, warning and mitigating the impacts a hurricane will bring when it interacts with human life.

WHAT IS THE FORECAST FOR A NUCLEAR WINTER?

During the late 1970s and the early 1980s, during the height of the Cold War, there were geopolitical tensions between the USSR (the Eastern Bloc) and the United States with its Western European allies (the Western Bloc), and there was a danger of nuclear war. Each side had an arsenal of nuclear weapons and there was a constant threat of using them on each other.

Science and history show us that if we drop a nuclear bomb on a city, the immediate consequences are catastrophic. Nuclear bombs are essentially bringing a piece of the Sun to Earth for a fraction of a second. The blast and heat of a nuclear bomb destroys buildings and kills all life in the immediate vicinity straightaway, followed by what is known as nuclear fallout, with radiation that would affect every living thing for miles around the radius of the blast and to smaller effect around the globe through the wind patterns. Despite knowing the devastation such a weapon could reap, no one had really thought about the long-term climatic effects of nuclear war. By this point, you may already be conjuring up images in your mind of a planet with deserted towns and cities in a dark and cold environment. TV shows and films were indeed produced in the early 1980s depicting post-apocalyptic futures after the breakout of a nuclear war, but they do not give us the complete picture.

What Has Weather Got to Do With It?

The idea of a nuclear winter came about in the early 1980s when there was an American Geophysical Union meeting to discuss the climatic effects of nuclear war. For a moment, forget the initial blast and radiation you would get if a nuclear bomb landed on a city. During the aftermath, you would have large fires burning all over the place and those fires would bellow out a huge amount of smoke into the atmosphere, where a small fraction of it would most likely make its way into the lower stratosphere. As you don't get rain clouds in the stratosphere, there is no process for the smoke particles to get 'rained out' and the smoke would stay in the stratosphere for a long time. Why is this important? Well, it was suggested that even a small amount of smoke in the upper atmosphere would decrease the amount of solar energy, i.e. sunlight, reaching Earth's surface, thereby reducing the global temperature. The physics is that simple.

Climate scientists who were looking primarily at the effects of volcanic eruptions during this time thought that they could use their developed climate models to help research the effects of a nuclear winter. The models they were using involved adding a load of particulate and sulphuric acid into the atmosphere to simulate a volcanic eruption, so they simply substituted that with the type of smoke that you would get with large fires.

The India-Pakistan Model

One of the most striking computer simulations of a nuclear winter scenario was performed by a group of scientists in the United States, who looked at the consequences of India and Pakistan entering into a nuclear war with each other. In the early 1980s, they each had about 50 Hiroshima-type atomic bombs so the model looked at how much smoke would be produced from fires if they went into all-out war. They predicted that 5 million tonnes of smoke would be injected into the atmosphere, which would gradually affect the whole planet. There would be regional differences in how temperatures would react, but landlocked countries would see temperatures below freezing. It would be darker and, due to the ozone being destroyed, the amount of UV light reaching Earth's surface would be much higher. Plant life, wildlife and natural ecosystems would all be affected. A lot of people were starting to take notice of the global consequences of starting a nuclear war and were realising that it would almost be suicide for the county starting it, let alone annihilation for the target.

But how do we know that the models are correct? Let's go back to volcanoes as the scientists themselves started there. In April 1815, Mount Tambora in Indonesia had a major eruption and at the time it was the most devastating eruption in recorded history. While estimates vary, the death toll was said to be around 70,000. With a huge amount of ash being pumped into the atmosphere and transported around Earth, it had a significant effect on the climate. By the following year, the global temperature had dropped by 0.4–0.7°C and while that doesn't sound a lot, it had an enormous impact and was known as the Year Without Summer. It was an agricultural disaster that meant major food shortages across the Northern Hemisphere. While the Year Without Summer was the result of a natural disaster, the basic principle of a large amount of smoke, ash and sulphur dioxide being released into the atmosphere blocking out the sunlight and reducing the global temperature is the same principle that is applied to a nuclear winter.

Despite the tensions of the Cold War, American and Russian scientists in the early 1980s presented the results of their nuclear winter scenarios to the then USSR and US presidents Gorbachev and Reagan. They listened and took on board the research with Ronald Reagan, even saying the mass use of nuclear weapons would wipe out Earth as we know it. As a result, there was a reduction in nuclear weapons on both sides, with several treaties being signed to commit both countries to nuclear reduction. The number of nuclear weapons across the world continues to fall, but unfortunately, at the same time there are more countries who own nuclear weapons. Recent studies using more advanced climate models than those used in the early 1980s still come up with the same results: the problem has not gone away.

COULD CLIMATE MODIFICATION SAVE THE PLANET?

Weather modification techniques, like rain enhancement, are helping some countries to combat water stresses on relatively small scales, although there are some more significant operations, as we've noted with China's current activity. There have been other attempts to modify the weather, such as hurricanes, hail storms and fog dispersal, but we are now entering a period where climate change is being felt on a global scale. Scientists say that time is running out to reduce emissions of carbon dioxide and keep global temperatures below a limit where we may restrict the long-term damage global warming could bring. We are relying on governments to change policy and act together to save the planet. Some see this as not enough, so are turning to the engineers. Are there any hard engineering techniques we can use to stop global temperature rising to levels that would bring irreversible damage to the planet?

To stop the global temperature from rising or even to push it down, there are some options being talked about. Some of the craziest suggestions have been to move Earth away from the Sun and to place large mirrors in space to deflect solar radiation, but both seem like ideas from a science fiction book. Still, some of the principles behind putting gigantic mirrors into space are not that fanciful. Clouds in the atmosphere are the natural reflectors of solar radiation coming down to Earth. The whiter and brighter the clouds are, the more reflective they become, eventually lowering the temperature on Earth. One project has been looking at spraying a fine mist of seawater into clouds in a marine environment. This could work especially in polar regions to cool the poles and help reduce the rate of sea level rise. There are no toxic chemicals required and it is seen as relatively affordable. However, it could severely mess up regional weather patterns and actually lead to a reduction of rain in some areas, and given this, it is a controversial type of weather modification, more needs to be done to promote any advantages over the disadvantages.

Another idea to reduce the global temperature is to mimic what happens naturally when a volcano erupts. We've seen very clearly throughout history that big volcanic eruptions temporarily reduce the global temperature. The 1991 major eruption of Mount Pinatubo in the Philippines sent huge amounts of dust, ash and aerosols high into the stratosphere, which resulted in a cooling of the global temperature by 0.5°C over the subsequent two years. That may not seem like very much but when we are talking about changes in global temperature, it is 1.5°C above pre-industrial levels that we need to limit our warming to, therefore a 0.5°C reduction is significant. The question is, could we artificially pump aerosols into the atmosphere that would have the same effect and reduce the global temperature? In theory, yes, we could release sulphur dioxide into the stratosphere, where it would form aerosols that reflect sunlight back into space. And, yes, it would reduce the global temperature. However, is pumping more greenhouse gases into the atmosphere to offset another greenhouse gas without really knowing the full side effects really a good idea?

Probably the most sensible option for climate modification to date and one that is gaining momentum among engineers is direct air carbon capture. Carbon capture isn't new. In fact, there are a number of greenhouse gas emitting industrial units that have carbon capture technology that removes carbon dioxide from the emissions before it gets released into the atmosphere. The carbon dioxide is then either stored in the ground or sold on for other uses. While this is a good step to reduce carbon dioxide emissions, it is only making the burning of fossil fuels carbon neutral as no net CO_2 is going into the atmosphere. This will only slow the inevitable rise in global temperature. What we really need is a carbon negative process that starts to reduce the amount of carbon dioxide in the atmosphere. This could soon be the reality, with technology being introduced that, explained very simply, uses big fans that suck air in, take out the carbon dioxide and release the clean air. It works on a small scale but to make serious inroads to reducing CO_2, we need to scale up this technology.

CLIMATE MODIFICATION FOR EVERYONE THEN?

After the Environmental Modification Convention was set up in 1978, in reaction to the clandestine Operation Popeye, it was decided that countries aren't allowed to 'militarize the weather'. With the addition of another convention on Biological Diversity in 2010, there was also a ban on other forms of weather modification or geoengineering.

How is this important? Well, essentially these conventions prohibit environmental modification techniques that change 'through deliberate manipulation of natural processes – the dynamics, composition or structure of Earth, including its biota, lithosphere, hydrosphere and atmosphere, or of outer space.' Options such as the cloud brightening project fall foul of these international conventions as there are possible climatic impacts on other countries and that would be a political hot potato. Any serious consideration of a hard climate modification technique on a global scale would need total international cooperation. That shouldn't be too hard, should it?!

WAR AND WEATHER

Any military commander will tell you that the weather has an effect on war, and there are certainly many examples throughout history that testify to the importance of knowing the weather and environmental conditions you are sending troops into. From 1588, when favourable winds and gales contributed to the British defeat of the Spanish Armada, to World War I, when it was all quiet on the Western Front during the 'mud season'.

WEATHER FORECASTS AND WAR

Official weather forecasts go all the way back to 1861 when the Royal Navy's Vice-Admiral Robert FitzRoy published the first daily weather forecast in *The Times* newspaper. FitzRoy was convinced he could forecast the weather and as a pioneering meteorologist, set up the forerunner of the UK's Met Office as part of the Board of Trade in 1854. The main purpose of the earliest weather forecasts was to give shipping and maritime some warning of impending bad weather. The early weather forecasts were very rudimental and FitzRoy was often mocked for getting them wrong. After a number of years with varying success in forecasting the weather he retired in 1865, but his early ideas started a movement of meteorologists seeking to gain more knowledge about how the atmosphere works and spawned further attempts to forecast conditions up to a day ahead.

Over the decades since FitzRoy's first forecast in *The Times*, meteorology developed as a science and weather forecasts became commonplace across Europe and North America. During the opening phases of World War I, the British Army soon came to realise they needed some assistance from the Met Office, particularly in predicting wind directions and strengths. As gas was being used on the battlefield, the Army needed to know which direction the gas would travel: predicting the winds higher in the atmosphere would help the artillery battalions and the Royal Flying Corp, who needed information on cloud and fog.

Meteorological information became so important to the military that more and more information was being requested on a daily basis. By 1918, weather forecasting had become an essential part of British military strategy. As technology and the military developed with the addition of the aircraft being used by the Royal Air Force, weather forecasts became an essential part of combat strategy. By the start of World War II, the Met Office was supporting the military with weather forecasts being provided by a few thousand staff in uniform.

Winning Weather Forecasts

Leading up to the D-Day landings in June 1944, there was meticulous planning surrounding all the military aspects of pulling off such an operation. Transporting thousands of soldiers in boats and aircraft across the English Channel was no mean feat, but one thing Allied commanders couldn't plan was the weather. Conditions had to be just right in the Channel. There could be no strong winds, heavy seas, low cloud or poor visibility. The landings had to take place at first light at low spring tide (to expose the German defences) and ideally, during a full Moon (to improve the light during a night-time crossing). All this meant that they had a window of three to five days in June to complete the operation.

Given this narrow opportunity, the Allied commanders advocated for a meteorological unit to give them as much information as possible during the planning phases of operations. That unit was led by Group Captain James Stagg, who had to look after three teams of forecasters from the Met Office, the British Navy and an American unit, who were all using slightly different forecasting methods. These differences meant that Stagg would often have to interpret three conflicting weather forecasts before briefing the Allied high commanders.

Additionally, he was being asked to provide weather forecasts up to five days ahead of time when 12–24 hours was regarded as the limit of predictions in those days. Weather observations were made from land, ship and aircraft positioned throughout the UK and France. One sticking point was that the German weather observations were coded, making it difficult to know what was happening on the ground. Thankfully, the codebreakers at Bletchley

Park in Milton Keynes were able to crack the code and the Allies could gain a better picture of the weather situation across Europe to use in their forecasts.

As the three- to five-day window of ideal lunar and tidal conditions got closer, Stagg was under huge pressure and the whole operation began to hinge on the weather forecast. Unfortunately, the conditions in late May and early June of 1944 were very unsettled, with weather systems bringing in strong winds and rain, which made the job of forecasting a suitable weather window very tricky. Eventually, 5th June was picked as a provisional date to launch D-Day (also known as Operation Overlord), but bad weather on the 4th looked set to continue so it was postponed. Every day the Allies delayed, the tides would be getting worse and the pressure on the meteorologists mounted.

Conflict within the forecasting unit continued with each of the teams coming up with different variations of a forecast. Stagg was pressurised by his superiors but couldn't change the uncertainties in the forecast – it seemed like an impossible job. The Met Office and Royal Navy forecasters were proposing that there might be a window of finer weather on 6th June, but the Americans disagreed. Stagg presented his forecast and suggested that the conditions would just be bearable. General Eisenhower, the Supreme Allied Commander, gave the instruction to go.

The forecast didn't quite go to plan as while the weather did settle down on the 6th, it was still very windy, with rough seas across the Channel. Many soldiers were seasick, boats were overturned and paratroopers missed their targets because of a stronger-than-predicted wind. However, as history tells us, the D-Day landings and Operation Overlord were a success. Had Group Captain Stagg not made the decision to give a forecast of finer weather that day, the next opportunity when the tides and Moon were ideal was two weeks later. Had they postponed the operation until then, the weather would have been far worse, with near gale winds and 20ft seas in the Channel. After this potent storm, General Eisenhower wrote to Stagg and said, 'Thank the Gods of war we went when we did'.

MILITARY METEOROLOGISTS

It was now clear how important weather forecasters could be in combat situations, but this importance didn't just apply to the Army. As the Royal Air Force expanded in World War II, it also became apparent that aircraft were highly susceptible to the weather and a RAF Volunteer Reserve (RAFVR) Meteorological Branch was set up. Thousands of Met Office personnel signed up and were active at military bases all over the UK and Europe, providing those vital forecasts. Thousands of lives were saved by knowing when flying conditions weren't going to be suitable.

Soon after the war ended, many of the uniformed meteorologists were disbanded, with only 200 retained to support weekend flying and participate in military exercises. With peacetime and no real need of a full unit of meteorologists, the branch was eventually disbanded completely. During the increasing intensity of the Cold War in the early 1960s there was once again a growing need for weather forecasts. In 1963, the Meteorological Branch was set up again but was formed into an official unit of the Royal Air Force known as the Mobile Met Unit (MMU). Their remit was defined to 'provide met[eorological] support to deployed UK armed forces at locations where local met support is insufficient'.

The A-Team of Meteorology

While the MMU was a full unit of the RAF, it was largely comprised of forecasters, engineers and observers who were reservists with civilian jobs within the Met Office, most normally at a military base. When they were called up into service, they were doing the same job but in uniform at a deployed location.

After providing support throughout the Cold War, along with various exercises, the real benefit of the MMU was exhibited during the Falklands War in 1982. They were deployed to the Ascension Island, where they provided meteorological support to the UK forces, particularly giving route forecasts for aircraft travelling down to the Falklands and for the Vulcan bombing runs on Port Stanley airfield.

In the 1990s, tensions in the Middle East escalated with the Gulf Wars. This meant the armed services deploying to another location where local meteorological support was non-existent and so the MMU team was also sent to the heart of the action. Originally, military commanders suggested they wouldn't need weather forecasts as it was 'always hot and sunny in the desert', but after a number of dust and thunderstorms, the MMU were very quickly sent out. At their peak, in the late 1990s and into the late 2000s, MMU teams were spread right across Eastern Europe, Iraq, Afghanistan and the Gulf, giving meteorological information direct to combat troops.

Go/No-Go Decisions

Now, the MMU forecaster is an integral part of the planning and operational process of a mission. Their weather forecast will include far greater detail than that you'd find on the TV, radio or online, with much more precise information about the height and amount of cloud, the visibility, how the wind varies throughout a forecast period and the onset of any severe weather, to name a few of the issues they cover. In fact, to highlight just how important the weather is, during daily military briefings the forecaster is given top billing. They inform the top commanders of the scope of weather they are likely to see in the day ahead, then specialised weather briefings to army and air force units continue throughout the day during operations.

ALL AT SEA: THE WEATHER AND THE NAVY

While the MMU are the weather forecasters for the British Army and the Royal Air Force, the weather is also vitally important to the Royal Navy as we've seen throughout history, stretching right back to the pioneer of weather forecasting, Vice-Admiral Robert FitzRoy. In the early years, forecasting for the Navy was the principal job of the Met Office, but as meteorology became even more important to nautical military operations a dedicated Naval Branch of Meteorology and Hydrography was set up in 1937. From this time, you could join the Navy and specialise in meteorology and hydrography (known as METOCs) in the Navy Meteorological Services. As with the MMU, the Navy METOCs play an integral role in day-to-day operations around the world, helping the service to exploit the weather to gain a military advantage. The MMU and Navy METOCs work closely together at the Joint Operation Meteorological Oceanographic Centre (JOMOC) in High Wycombe, UK.

CAN THE WEATHER BECOME THE ULTIMATE MILITARY WEAPON?

Irving Krick, an American meteorologist and engineer, once said, 'The nation that first acquires control of the weather shall be the leading power in the world.' We've already seen how weather is crucial for the military to operate at their full capability, so it may not be surprising to learn that governments throughout history have attempted to manipulate the weather either to benefit their own operations or to hinder the enemy.

Let's start with manipulation of weather to maximise one's own operations. In 1942, during World War II, many flying hours were lost due to foggy airfields in Britain. Winston Churchill ordered his chief scientific advisor, Lord Cherwell, to come up with a solution with the 'utmost urgency'. Lord Cherwell's response was to light multiple fire pits along the side of airport runways in order to raise the temperature of the air around the landing strip and burn away the fog. Named FIDO (Fog Investigation and Dispersal Operation), the first successful experiment occurred in November 1942, when 200 yards of dense fog was cleared. After this, FIDO units were installed at a number of English RAF stations. One of the major issues with FIDO was the cost. The operation needed a huge amount of fuel – 100,000 gallons of petrol and kerosene per hour. At the height of the war's air operations, it was deemed value for money as it enabled many more flying hours and ensured a huge reduction in aircraft lost to crashes while trying to land in the fog. It was even reported at the time that the use of FIDO shortened the war and saved the lives of thousands of airmen.

In a secret operation during the Vietnam War, the US government used weather modification techniques to try and gain the upper hand. In Southeast Asia, the US initiated highly classified cloud-seeding procedures in an operation named Popeye, the aim of which was to increase rainfall in areas around the Ho Chi Minh Trail to prevent the Vietnamese army using their supply trucks on the roads. The idea was that the increased rainfall would soften the road surface, cause landslides and local flooding, and wash out parts of the supply route. The cloud-seeding operation was conducted by the 54th Weather Reconnaissance Squadron, whose normal role was to gather weather data. They used the slogan 'make mud, not war'. Two sorties of cloud seeding were undertaken every day during the rainy season (March to November) for five years. It is not officially known how successful Operation Popeye was, mainly because of the secrecy surrounding it. After word got out about the operation there were multiple hearings in US Congress, but military commanders and President Nixon denied it even existed.

Nevertheless, there was a huge public outcry surrounding Operation Popeye. The use of weather modification as a military weapon was seen as setting a very dangerous precedent. In response, an international treaty, the Environmental Modification Convention (ENMOD), came into force during 1978 and was designed to ban the use of weather modification techniques for the purpose of inducing damage or destruction. Since then, weather modification research has been limited. There are plenty of conspiracy theories out there that suggest clandestine projects still exist to manipulate the weather for the military advantage.

HAS THE WEATHER EVER STARTED A WAR?

In answering this question it's important to remember the fundamental difference between the weather as the day-to-day and week-to-week atmospheric conditions we experience and the climate being the much longer-term average conditions over decades.

The changing weather from day to day or over weeks to a month hasn't directly led to any major war breaking out across the world as far as we're aware, although it's possible climate change may have done. We do need to be a little careful here directly linking climate change with conflict around the world, but it is certainly likely to be a contributing factor. In 2018, the World Bank issued a report in which they stated that climate change adversely affects food production, particularly in rural areas, where 80% of the extreme poor live. To simplify this, imagine if the area you live in enters a severe drought and your food production is no longer sustainable. You would then seek to move. Essentially, this pattern can underpin migration from country to country which, in turn, may lead to an increase in conflict. The same can be said about more severe weather events such as storms, flooding and extreme heat waves. Some researchers have suggested that there could be a 10–20% increase in conflict for every 0.5°C increase in global warming.

One example of this type of 'climate conflict' is in Syria when a drought lasting five years from 2006 to 2011 caused 75% of Syria's farms to fail and 85% of livestock to die, according to a UN report. This drought displaced over 10 million Syrians and the country spiralled into a vicious civil war. It is thought that droughts in Yemen, Libya and South Sudan have all exacerbated the civil wars.

Climate Change Refugees

In the 2018 World Bank report, the issue of preparing for internal climate migration was explored and it was suggested that by 2050, if no action on climate change is taken, there will be more than 143 million internal climate migrants across Sub-Saharan Africa, South Asia and Latin America. The poorest people will be forced to move due to decreasing crop production, shortage of water and rising sea levels. We're not just talking about developing countries either: the developed world will also suffer.

The 2018 Paris Agreement advocates the need to keep the global temperature increase to below 1.5°C over pre-industrial levels, but it is thought that this figure is likely to be exceeded. Some estimates suggest that even if we keep to this figure, the global sea level is likely to rise by 50cm by 2100 (and this is a rather conservative figure compared to some climate models we've seen). This would mean major cities such as New York, Miami, London, Bangkok and Shanghai would suffer significant coastal flooding. There would be over 500 global cities facing this problem. Many of the people living in these cities would be forced to move inland. In a warmer world, deserts would start to move northwards so that much of the Mediterranean would become a desert region. Unless there is mitigation to counter the impacts, you are likely to see climate migration from those areas further north. While much of this climate migration isn't going to directly lead to conflict, as we have seen in history, it could lead to instability and pose a threat to peace.

CLIMATE
CHANGE

NATURE'S CHANGING CLIMATE

Unpicking Earth's past climate is a mammoth task. The subject is immense. To understand present and future changes in climate, it's essential to first clarify past climatic rhythms – the long period of time before 'human-induced' climate change became the most influential driver in this subject.

Past Climate in a Nutshell

Earth has always undergone changes in climate when the mean global temperature has slipped low enough for ice ages and has risen high enough for warmer spells, when even the poles have been ice-free.

Over the past billion or so years, there have been about five or six major glacial periods with brief, warmer interludes or interglacial periods. The earliest of these was about 2.4 to 2.1 billion years ago, and the most recent peaked around 18,000 years ago. During this period, the average global temperature was six degrees lower than the mean global temperature today.

Six degrees may not seem like a large temperature range, but here the global temperature represents the planet's whole surface, from the North Pole to the equator to the South Pole, and everything in between. It's an average of all seasons, winter/summer, dry/wet, cold/hot. In climate terms, a change of six degrees is colossal.

During the last glacial period, the slow drop in temperature allowed ice sheets to grow and spread from the colder reaches of the poles towards the mid-latitudes. Freshwater became locked as ice. As a result, sea levels fell significantly, and there was less available water/water vapour in the hydrological cycle. Glacial periods are much drier; beyond the extent of ice, the land suffers far more from drought and desertification. The mechanisms that, under current conditions, drive global weather are weaker, including convection, evaporation and larger-scale distribution of water vapour through strong winds as well as weather systems generally.

There is also convincing evidence that the weakening of the major ocean currents contributes to climate shifts. The warm ocean currents that transport warmth northwards and send cold, deep water southwards fail during cold global phases.

What Comes First?

Within an ice age, there are significant fluctuations. It's a far more complex system than just the atmosphere becoming colder. Within a glacial period, there are mini peaks and dips in global temperature. The glacial periods that have dominated Earth so far have been mostly due to natural variations in solar radiation.

Among so many variables, the systematic rising and

setting of the Sun is a known constant when it comes to climate and weather, but the Sun's radiation, long-term, does vary in a number of ways. The Sun's intensity changes over long periods of time. It has profound effects on the Earth's climate and, ultimately, on life and biodiversity. The Sun's solar output varies in an 11-year cycle; this, in turn, reinforces any cooling or warming when other processes come together.

The shape of Earth's orbit, or eccentricity, also varies through time. Every 100,000 years, it goes from near-circular to more elliptical, before returning to a near-circular orbit. During the more elliptical orbits, Earth is further away from the Sun, which at certain times of the year results in cooler spells. Currently, Earth is on its elliptical path and is furthest away during the Northern Hemisphere's summer.

Earth's tilt, or obliquity, changes over 41,000 years, from 22.1 to 24.5 degrees – it's currently at 23.4 degrees. The larger the tilt, the more extreme the seasons, leading to harsh winters and hotter summers. This also contributes to Earth's climate and has a greater influence on climate than the planet's distance from the Sun. Solar radiation is also affected by how much the Earth wobbles on its axis. Like a spinning top that can develop a 'wobble' – this is called precession. The precession is due to the battle of gravitational pull from the Sun and Moon on Earth.

When these different aspects of Earth's movement coincide, the level of solar radiation can weaken significantly, and Earth's climate responds. However, there are other processes at work as well. Positive feedback mechanisms play a part in enhancing the effects of cooling, or heating, once climate change is heading in one direction.

For example, an increase in ice over a vast area increases the albedo, and there is more reflection of sunlight and less absorption of light (and thus less heat); the environment cools down further. Although systems like this can be erratic, Earth's history is not without spells of accelerated cooling that have reinforced the growth and extent of ice.

Another mechanism that has been evident during an ice age is falling levels of carbon dioxide (CO_2). This is mainly due to absorption by the seas and oceans. CO_2 is a powerful greenhouse gas that traps heat and keeps the lower atmosphere warmer than if it were not present. The trapping of CO_2 during extended and intense cold spells inhibits heating and allows for the global temperature to fall further. But once this CO_2 is released, accelerated warming happens and the climate shifts again.

These changes are referred to as 'tipping points' whereby large climatic patterns shift so much that once a point has passed, there is no return and the consequences are huge to the climate we currently know.

Although the risk of reaching a tipping point isn't fully understood, it is considered a low probability by the end of this century. Nevertheless, the consequences of reaching a tipping point are so severe that concern has been expressed among some scientists.

Volcanic eruptions present another compelling example of natural climate changes, though their effects operate on different timescales. Major volcanic events inject vast quantities of ash and gas into the upper atmosphere, where these materials can persist for months or even years. This volcanic veil acts as a solar filter, reducing the amount of incoming sunlight and creating a cooling effect across the planet. The most recent example of this occurred after the 1991 eruption of Mount Pinatubo in the Philippines. After releasing millions of tonnes of sunlight-reflecting ash and gas high into the atmosphere, the global average temperature fell by 0.5 degrees the following year.

In summary, natural changes in solar radiation and other natural terrestrial processes have significant effects on the climate system, alternately plunging it into the realms of an icy existence and flipping it to a warmer environment.

The spatial and temporal scale on which this happens is almost unimaginable. Earth's climate history is fundamental to understanding what will happen next. Against a background of 2 billion years of global temperature falls and rises – changes taking place over tens of thousands of years within certain temperature bounds – imagine an accelerated rise in the last 100 years that is unprecedented over the past 2 billion years: an injection of heat so rapid it disturbs everything.

This is human-induced climate change, and it is already changing weather patterns.

Since the Industrial Revolution, carbon dioxide emissions have been steadily increasing, and the rate of increase has been accelerating. As a result, the global temperature is spiralling, and Earth will continue to heat up until radical action is taken to stop further emissions and ultimately to reduce the amount of carbon dioxide in our atmosphere.

Climatologists are clear that as the global temperature rises, the frequency and intensity of extreme weather will continue to increase. While weather and climate are obviously linked, there is an important distinction between the two.

Weather refers to the current and near-future conditions, from day to day up to months and even seasons ahead, whereas climate describes the average weather conditions over decades or trends in weather patterns. This means that there can be large variations in the weather on a daily or weekly basis, including extremes, that are smoothed out over a longer period of time and become less significant.

Or to put this another way ... your mood today (weather) doesn't tell me about your personality (climate).

Whenever there is extreme weather such as a heatwave, flooding or a powerful storm, this question is almost guaranteed to be asked:

'Is This Heatwave/Flood/Storm Due to Climate Change?'

Scientists employ a sophisticated method called attribution studies to determine whether climate change has increased the *likelihood* of severe weather events. This process involves running climate computer models with real-time weather data, then comparing these results against simulated scenarios where the atmosphere contains pre-industrial levels of carbon dioxide. By comparing these sets of model runs – one reflecting our current carbon-rich atmosphere and another mimicking conditions before widespread industrialisation – researchers can calculate how much more probable an extreme weather event has become due to human influence on the climate.

A UK heatwave in 2022 provides a compelling case study using this methodology. On 19th July 2022, the UK saw a temperature exceeding 40°C for the first time in recorded history. Attribution studies revealed that while such extreme temperatures would naturally occur perhaps once every thousand years, human-caused climate change had dramatically increased the likelihood of this happening, making the event ten times more likely to happen. What should have been an extraordinarily rare occurrence – the kind of weather that might visit a region only once in a millennium – had been transformed by our warming atmosphere into something far more probable, illustrating how climate change is fundamentally rewriting the rules of what we can expect from our weather systems.

Attribution studies can be conducted on all manner of weather events, from storms, droughts and wildfires to high and low temperatures.

We are witnessing an unprecedented increase in extreme weather across every part of the planet. Every skewed rhythm, every changing pattern that plays out in our atmosphere is influenced by the way human beings currently live on Earth.

MILITARY METEOROLOGISTS

What's the Difference Between Climate Change and Global Warming?

These terms seem interchangeable. However, global warming is one of many consequences of climate change. Climate change encompasses human-induced alterations to the climate (including the seas) as well as the natural variability of Earth's atmospheric patterns, such as El Niño and La Niña. It also includes the impacts of warming, such as ice melt, sea level rise and extreme weather events. Global warming is a key aspect of the climate change issue. It specifically refers to the warming of the climate due to an increase in greenhouse gases (GHGs) – gases that are now far higher in concentration in the atmosphere because of human activities such as burning fossil fuels, as well as industrial and agricultural practices.

The Greenhouse Effect

The greenhouse effect and greenhouse gases are synonymous with global warming. The purpose of a greenhouse is to allow maximum sunlight into a closed environment, where it is transformed into heat, which is then trapped, thereby raising the temperature inside the greenhouse to a higher level than it would be otherwise. Greenhouse gases trap heat and then re-emit that heat.

The Greenhouse Gas (GHG) line-up:
- Water vapour
- Carbon dioxide
- Methane
- Ozone
- Nitrous oxide

How Do Greenhouse Gases Trap Heat?

Light energy travels to Earth and is absorbed by Earth's surface, where it transforms into heat energy. This heat energy is then re-emitted into the lower atmosphere. Some escapes. However, greenhouse gases capture much of the heat, which is then absorbed and re-emitted. The presence of greenhouse gases across the planet means this process is ongoing, and the heat remains close to Earth's surface.

The reason these gases trap heat is due to their molecular structure. This particular set of gases has a loose molecular structure, which allows them to vibrate when energy (in the form of heat) is absorbed. This infrared energy energises the gas, which is then re-emitted.

Earth Without the Greenhouse Effect?

Greenhouse gases are essential for sustaining life on Earth. Without them, Earth's global temperature would be more like minus 18°C, instead of a more comfortable 15°C. Greenhouse gases are the blood that flows around the body, keeping it warm and alive. The level of concentration of these incredible gases is a fine line between too much and too little. But how is this concentration rated? Each greenhouse gas is rated on three characteristics:

1. Their abundance in the atmosphere;
2. How long they remain in the atmosphere;
3. Their impact when they are in the atmosphere (their global warming potential).

All greenhouse gases trap heat. Some are more potent than others. For example, methane is many times more potent as a greenhouse gas than carbon dioxide, so it is initially far more devastating to the climate due to its effective heat absorption. It's a by-product of burning natural gas, but also produced in significant volumes from livestock farming and released in substantial quantities from melting permafrost. However, it doesn't linger as long in the atmosphere as carbon dioxide, but like carbon dioxide, levels of methane continue to rise.

What Is the Carbon Cycle?

Carbon dioxide is the most abundant greenhouse gas after water vapour, and for good reason: it is part of the breath of life, being continually inhaled and exhaled on land and sea by animals, plants, trees, plankton, rocks and soils.

Just like Earth's water cycle, the planet's carbon cycle transports, absorbs, emits and redistributes carbon dioxide. This is known as a biogeochemical cycle. During photosynthesis, plants absorb carbon dioxide and, along with sunlight and water, form carbohydrates. This is stored within a plant and helps it grow, but the waste product of this process is oxygen, which is released into the atmosphere.

Nature is superbly efficient in utilising the surrounding energy. Green plants contribute significantly to the oxygen we breathe, but hidden beneath the ocean's surface, phytoplankton and cyanobacteria are thought to produce up to half of the world's oxygen.

Buried organic matter stores huge reservoirs of carbon in the form of coal, gas and oil. We have foraged these fuels for thousands of years, and when foraging becomes a global industrial obsession, the volumes of carbon dioxide that then find their way into the atmosphere tip Nature's balance from the long-term norm to something unprecedented.

CO_2 Measurement

Carbon dioxide levels in our atmosphere are measured in parts per million (ppm) – a simple way of counting how many CO_2 molecules exist among every million molecules of dry air. When scientists say we've reached 430 ppm, for example, they mean that out of every million air molecules, 430 are carbon dioxide.

These measurements are incredibly important in understanding the future health of our planet. The rise in carbon dioxide levels since the Industrial Revolution is correlated with the rise in global average temperature. Climate scientists feed this data into sophisticated computer climate models and analyse how Earth's natural systems unfold.

Over the last 800,000 years, CO_2 levels have fluctuated

naturally between 180 ppm during ice ages and 280 ppm in warmer interglacial periods. Before industrialisation, levels held steady at around 280 ppm.

In 1958, Dr Charles Keeling began measuring carbon dioxide from Mauna Loa volcano in Hawaii – chosen for its isolation from pollution sources. This remote station has become the gold standard for CO_2 monitoring, providing consistent daily readings that continue to this day.

The resulting data, known as the Keeling Curve, has become the benchmark for tracking greenhouse gas increases over seven decades, revealing the true pace of our atmospheric transformation.

The Keeling Curve reveals a relentless year-on-year rise in CO_2, accelerating in recent decades. These detailed observations also capture Earth's seasonal breathing – concentrations of CO_2 naturally dip each spring and summer as Northern Hemisphere vegetation absorbs CO_2 through photosynthesis, then peak in autumn and winter when plants and soils release it back through respiration.

Keeling was among the first to link rising CO_2 directly to fossil fuel combustion. When his measurements began at 316 ppm – already significantly above pre-industrial levels of 280 ppm – he issued a prescient warning: if concentrations exceeded 400 ppm, the world would face a clear and present danger from runaway warming.

In 2013, Mauna Loa recorded a CO_2 level of 400 ppm for the first time. Today, this threshold may seem low, given that concentration levels are rising faster than ever.

Earth has endured such carbon loading before. Fifty-five million years ago, a sudden release of CO_2 and methane triggered over 5°C of global warming in a geological instant. The planet became a hothouse – widespread extinctions swept land and sea, most regions turned uninhabitable and Arctic waters warmed to subtropical temperatures.

Recently, remains of trees have even been unearthed 300 miles from the South Pole. This was the Paleocene-Eocene Thermal Maximum or PETM. It lasted about 100,000 years. It may seem a far-flung, distant event, before the advent of *Homo sapiens*, but it provides a stark reality of what is happening right now. The warming during this event occurred over a span of 20,000 years. Today, CO_2 emissions have escalated over just 200 years. Data from this period in Earth's climatic history provides more than just important lessons about our road ahead; it also signposts the way Earth's response can nonchalantly flip environments.

What Is Climate Sensitivity?

At the heart of climate science is the sensitivity of Earth's climate to increases in carbon dioxide. This measurement indicates the extent of global surface warming that is expected to occur in response to human-induced CO_2 emissions. Studies on climate sensitivity use climate models, recent observations and data from Earth's past climate to estimate current and future temperature trends. The processes are complex and some level of uncertainty has to be considered. Numerous scenarios include feedback mechanisms such as how ice sheets will change, how cloudiness, including the types of cloud, change, the lag time in ocean response and the take-up of CO_2. This means that results produce a range of future global temperatures rather than a definitive number.

These projections are commonly described as 'emission scenarios'. At the most ambitious end, 'low emission scenarios' involve rapid, deep cuts in CO_2 emissions that could limit the global average temperature rise to around 1.5°C above pre-industrial levels – though achieving this requires global emissions to peak soon and fall drastically by 2030. The 'current policy scenario', based on existing government policies and current Nationally Determined Contributions (NDCs) under the Paris Agreement, puts the world on track for approximately 2.6–3°C of warming by 2100. Finally, 'high emission scenarios', where countries fail to implement significant CO_2 reductions, could lead to a warming of 3.5°C or higher by the end of the century.

What Are the Global Impacts of a Warming World?

Climate change research spans centuries, though recent decades have seen an explosion of studies across all aspects of this discipline, from science to impacts to solutions. In 1856, scientist Eunice Foote first described the greenhouse effect, concluding that carbon dioxide traps heat and warms our atmosphere. Forty years later, Swedish scientist Svante Arrhenius predicted that doubling atmospheric CO_2 would raise global temperatures by 5–6°C.

By the mid-20th century, concerns rippled through the scientific community. Early computer models calculated that CO_2 doubling would produce 2°C of warming – without even considering feedback mechanisms, such as accelerating ice melt or the impact of deforestation, like that across the Amazon.

As computing power advanced, future scenarios came into sharp focus. Satellite imagery added a keen eye to Earth's responses: continual monitoring revealed rising temperatures, shrinking ice sheets, retreating glaciers, expanding deserts and a planet in a state of metamorphosis.

We have become witnesses to predicted climate change unfolding before our eyes in real time. Today's supercomputers can process climate data in hours that once took months, creating increasingly sophisticated models that reveal Earth's sensitivity to rising CO_2 levels.

But you really don't have to look that far to see how this plays out. Let's look at Mars and Venus.

Mars – too few greenhouse gases

Although Mars' thin atmosphere is mostly carbon dioxide, it lacks methane and water vapour, so it has a very weak greenhouse effect. Consequently, Mars has a mostly frozen surface, and so far, there is no evidence of life.

Venus – overloaded with greenhouse gases

Venus has an atmosphere dominated by CO_2, with 19,000 times more than Mars and 154,000 times more than Earth.

Even though Mercury is closest to the Sun, Venus is the hottest planet in our solar system. This is due to its runaway greenhouse effect!

Climate Science Today and How It Supports Action

The Intergovernmental Panel on Climate Change (IPCC) is a United Nations organisation that publishes regular Assessment Reports (AR) on Earth's future climate. It collates the work of thousands of scientists from around the world. Released every six to seven years, these comprehensive reviews synthesise the latest climate science, offering policymakers authoritative evidence on warming trends, projected impacts and mitigation pathways.

This information supports the annual meeting of the 'State Parties to the United Nations Framework Convention on Climate Change' (UNFCCC), which serves as the supreme decision-making body for global climate action. These yearly 'Conference of the Parties' or 'COP' meetings assess progress and negotiate future measures to address climate change.

Negotiators rely on IPCC findings to set emission reduction targets, establish adaptation strategies and justify the urgency of climate action. In essence, the IPCC provides the scientific compass that guides COP's political decisions.

The story of shifting climate targets reveals both progress and mounting urgency. In 2015, COP21 in Paris was hailed as a breakthrough when most nations committed to limiting global warming to 2°C with a goal to pursue efforts limiting warming to 1.5°C, based on the IPCC's Fifth Assessment Report (AR5).

Yet just three years later, a special IPCC report delivered sobering news: the difference between 1.5°C and 2°C of warming wasn't marginal – it was profound. A half-degree separation exists between extreme and catastrophic impacts, affecting millions more lives, ecosystems and communities.

This revelation transformed the climate conversation overnight. The 'successful' 2°C target suddenly seemed dangerously inadequate. Today, as global temperatures continue their relentless climb, we find ourselves in a race against time to stay below the 1.5°C threshold – a target that once seemed ambitious but now appears increasingly elusive.

The Sixth Assessment Report (AR6), published in 2023, is the most definitive yet.

AR6 chronicles a world transformed by the rise of greenhouse gas emissions. It concludes that human activity has unequivocally warmed our planet's atmosphere, oceans and land, unleashing widespread consequences across natural ecosystems and human societies. These impacts have already surpassed previous scientific predictions in both scope and intensity.

On our current trajectory, it is unlikely that the world will maintain a global temperature below the long-term average threshold of 1.5 degrees.

What Are the Main Weather Impacts?
Heatwaves, Droughts and Wildfires
Climate change is expected to bring more intense heatwaves that will be hotter and last longer in the future. In addition to its impacts on health, extreme heat results in excess deaths, and it may also lead to an increased risk of longer and more severe drought conditions in some areas.

The mid-latitudes in particular are projected to have a higher incidence of extremely hot days, during which the average daily temperature could be more than 3°C higher than it would be under current conditions. Importantly, with global warming exceeding 1.5°C, the risk of heatwaves and the frequency of extreme temperature rises increase further.

Towards the Poles, the air temperature is expected to rise by the greatest increment.

Alaska's 2019 wildfire season demonstrated the volatile extremes that climate change can unleash within a single year. Following an unusually wet spring, the state experienced its driest June on record – a dramatic shift that set the stage for catastrophe. Spring ice break-up occurred two months ahead of schedule in March rather than the typical May, signalling the unprecedented nature of the seasonal disruption. As the landscape dried to critical levels, wildfires erupted across the state. Throughout June and July, dense smoke advisories became a constant feature of daily life as fires burned out of control across vast areas of wilderness. The situation intensified dramatically when July temperatures soared 15 to 20 degrees above the monthly average, baking Anchorage and extensive regions of southern Alaska under relentless heat. This extreme warming was caused by a persistent high-pressure system known as a 'heat dome' – a blocking weather pattern that trapped hot, dry air over the region. Once established, this atmospheric configuration became self-perpetuating, allowing the dangerous combination of heat and aridity to maintain its grip on the landscape for weeks, transforming Alaska's normally temperate summer into a furnace that fed the advancing flames.

The devastating Los Angeles wildfires of January 2025 exemplified the dangerous phenomenon of climate whiplash, where extreme wet conditions rapidly give way to extreme drought. The previous winter's abundant rainfall had produced unprecedented vegetation growth across the region's hills and mountains, creating vast quantities of combustible fuel. However, by January 2025, the landscape had endured months of record-breaking drought and heat, with Los Angeles receiving less than a quarter of an inch of rain since autumn, leaving the lush vegetation completely desiccated.

When the Santa Ana winds arrived with brute force, gusting up to 100 miles per hour as they swept down from the Great Basin, they created the perfect conditions for spreading catastrophic wildfires.

The Palisades and Eaton fires alone destroyed over 18,000 structures, killed at least 30 people, and forced more than 200,000 residents to evacuate. Climate change has extended California's traditional fire season well into winter, creating a deadly overlap with the peak Santa Ana wind season that historically occurred when vegetation was naturally moist from winter rains.

The number of spikes in extreme weather over such a short period of time is astonishing. They are happening far more frequently than even a few decades ago.

An example of the increasing frequency of such impactful weather was laid out in a stark scientific paper authored by the Met Office's Dr Gillian Kay in 2025, where they analysed the probability of reaching 40 degrees in the UK. It was vanishingly small in the 1960s – but even by the 1980s it had risen to a 1-in-150-year event. Fast-forward to the UK summer of 2022, and the theoretical became real.

On 19th July 2022, the most extreme heatwave ever to hit the country smashed temperature records across the country, with Coningsby in Lincolnshire shattering the mercury at 40.3°C.

The UK government declared a national emergency, issuing the highest possible level of warning, with the risk of 'illness and death occurring among the fit and healthy'. There were five extreme hot spells that summer, and across England, this resulted in 2,985 excess deaths.

It was also Europe's hottest summer on record to date, triggering devastating heatwaves and over 61,000 heat-related deaths in 35 European countries. Nearly 900,000 hectares of land burned across the continent due to wildfires.

Based on the current climate, Dr Kay and her team

concluded that we are witnessing an accelerating and unstable trend. Within 12 years, the risk of exceeding 40 degrees will not only be 50–50 in the UK, but record temperatures could reach several degrees higher, where heatwaves extend over a month rather than a week.

Heavy Rain and Floods

In a warming world, the atmosphere can hold at least 7% more moisture for every °C of warming. This extra moisture means there is a potential for more intense or extreme rainfall.

The United Nations' Intergovernmental Panel on Climate Change (IPCC) Sixth Assessment Report concluded that as Earth continues to warm, it is 'very likely' we will experience heavier precipitation events across the world. The Met Office suggests that while summers will tend to become drier overall in the UK (average precipitation over June, July and August), the heaviest rain will be more intense, leading to a higher risk of flash flooding.

And it's flash flooding that can bring some of the most devastating impacts.

In Valencia, Spain, flash floods in late October 2024 were one of the worst natural disasters Europe has seen. A storm system stalled over eastern Spain with an entire year's worth of rain in just eight hours, triggering catastrophic flash floods that tore through towns and suburbs with terrifying speed. More than 200 people lost their lives, with 60,000 homes and 115,000 vehicles destroyed, making it the most expensive insured disaster in Spanish history.

The disaster was made worse by a combination of human and environmental factors. Many residents received flood warnings too late to escape safely. Flood defences, designed mainly to protect central Valencia, left rapidly growing suburbs more vulnerable.

Climate change also amplified this Spanish storm. Scientists noted that the Mediterranean Sea was unusually warm, fuelling the storm's intensity and making these types of events more frequent and powerful.

And in early July 2025, Central Texas was devastated by a sudden flash flood that turned into one of the deadliest summer disasters in decades. A fierce storm brought more than a foot of rain across Hill Country. The dry, rocky soil caused much of that rainfall to turn into raging run-off. In areas like Kerr County, the Guadalupe River surged dramatically, rising extremely rapidly, catching many asleep and completely unprepared.

At least 135 people, including 37 children from a summer camp – Camp Mystic – lost their lives.

This area was always prone to flash flooding, but after a lengthy dry spell, dry soil isn't able to absorb sudden rainfall as well as moist ground. This means that when a big storm hits with intense rainfall, the water runs off more rapidly into rivers, causing water levels to spike dramatically.

It may seem strange that predicted climate change repercussions bring both wetter and drier spells. It's the level of extremity of individual weather events that are of most concern; intense heatwaves causing prolonged drought conditions, or a storm's injected energy producing flooding from torrential rain.

This new reality of hydroclimate extremes – swinging violently from flood to drought – has brought what can only be described as a form of environmental hell upon vulnerable communities, transforming the natural cycles that once sustained life into forces of unprecedented destruction.

However, it's the succession of one extreme after another that can lead to greater destruction.

How Do Oceans Suffer From Heatwaves?

Higher global temperatures also affect the oceans. Recent heatwaves across the UK and Europe have extended to neighbouring offshore waters, where sea surface temperatures have risen accordingly. This alters the chemical, biological and physical dynamics of these seas. Increases in sea temperature reduce the oxygen content of the water and increase its acidity.

Coral bleaching is one very visible sign of excessive warming during the warmer El Niño years. Swathes of marine vegetation, coral and fish are wiped out. Severe coral bleaching affected the world's southernmost reef at Lord Howe Island, located off the coast of New South Wales, Australia. The bleaching occurred during the summer of 2018/19 when severe heatwaves hit much of the country. March 2019 was Australia's hottest March on record to that date; temperatures were 2 degrees higher than average. The most damage was observed in shallow waters. One solution for hardy marine life is to migrate to cooler waters, but this puts immense pressure on existing cooler ecosystems.

In 2023, an unprecedented surge in marine heatwaves occurred across Earth's oceans, setting new records in intensity, geographic extent and duration. Many of these heatwaves lasted well over a year, and 96% of the ocean surface was affected. These extreme warming events are elevating sea surface temperatures far beyond normal thresholds, triggering species and populations with narrow distributions, limited mobility and proximity to their warm distribution limits to face severe temperature stress.

The impacts ripple through marine ecosystems, forcing mass migrations as species flee to cooler waters and causing significant biodiversity loss when organisms cannot adapt quickly enough. Perhaps most concerning for long-term planetary stability, these persistent heatwaves are accelerating surface currents and disrupting the ocean's natural circulation patterns. Scientists warn that the reversal of ocean circulation in the southern hemisphere could double current atmospheric concentrations of CO_2 by compromising the ocean's crucial role as a carbon sink, creating a dangerous feedback loop that threatens to fundamentally alter global weather systems and accelerate climate change beyond current projections.

Sea Level Rise
Sea levels are expected to rise, but again, how high will depend on how much CO_2 emissions are curbed. The

response of the oceans and seas to global warming is slower than other land-based Earth systems, but the lag time means that even when warming slows and eventually is reversed, sea level rise will continue for a long time after this. The timeline to calculate future sea level rise extends beyond 2100. More than 600 million people (around 10% of the world's population) live in coastal areas that are less than 10 metres above sea level. Nearly 2.4 billion people (about 40% of the world's population) live within 100 kilometres (60 miles) of the coast.

Small islands are most vulnerable, with an invasion of saltwater poisoning delicate ecosystems and excessive flooding resulting in uninhabitable zones, and ultimately, much of the low-lying land being submerged underwater.

Ice Sheet Melt

The risk of losing major surface ice sheets over Greenland and Antarctica could increase sea water levels by metres, and these instabilities are now triggered at much lower warming levels than previously thought – with recent research suggesting even current warming poses significant risks. There is strong evidence that the Arctic is now warming nearly four times faster than the global average, with accelerated ice loss producing a positive feedback mechanism that adds even more warmth to high-latitude seas as darker, melted surfaces absorb solar energy previously reflected by white ice.

This is another 'tipping point' mentioned earlier where

the loss of sea ice could lead to runaway global warming with no chance of returning to a situation where Arctic sea ice would form over the winter, providing Earth's natural refrigeration system.

Species Risk
On land, the risk of species loss and extinction rises exponentially with added warming. A landmark 2018 study by the Tyndall Centre analysed 80,000 species across 35 wildlife-rich areas, finding that 50% could be lost without climate policy (at 4.5°C warming), reduced to 25% if warming is limited to 2°C.

More recent research reveals even steeper risks: a 2024 study shows that nearly 180,000 species could face extinction by 2100 at just 1.5°C warming, with risks accelerating dramatically beyond this threshold. These impacts include drought, erratic rainfall, water shortages, biodiversity loss and food chain disruption. Forest fires proliferate in arid conditions, invasive species like disease-carrying mosquitoes expand their ranges, and critical ecosystems like tundra and boreal forests face degradation, challenging the balance of these delicate environments that are essential for global climate stability.

What Are the Human Impacts?

According to a 2024 report from the World Meteorological Organisation, there were approximately 617 significant extreme weather events worldwide in 2024, of which 152 were classified as 'unprecedented'. These events displaced around 825,000 people and resulted in about 1,700 confirmed deaths.

At publication, the warmest year on record globally is 2024, where it was 0.12°C warmer than the previous record holder of 2023. This makes 2024 the first year to exceed 1.5°C above the pre-industrial level.

The ten hottest years globally have all occurred within the last decade.

Everyone is vulnerable to the health impacts associated with climate change, including life, livelihoods, water supply and food security. Some populations are disproportionately vulnerable: low-income communities, rural populations, people with disabilities or chronic medical conditions, immigrant groups, indigenous populations, children, pregnant women and the elderly. Additionally, those dependent on agriculture and fisheries, including small island states and developing countries, are also affected.

People die during heatwaves, from heat stress and heat stroke, affecting those with underlying health issues such as respiratory and cardiovascular diseases. During the 2003 UK heatwave, 2,000 people died, and in Europe the same heatwave claimed 70,000 lives.

It's not just the heat that has a detrimental effect on health. The quality of the air also deteriorates during such hot spells, so the number of people affected rises further. Low-level ozone, airborne particulates and an increase in pollen due to longer growing seasons are just some of the other factors that play into this quagmire of intensive heat. Vector-borne diseases, such as malaria and dengue fever from mosquitoes or Lyme disease from ticks, are influenced by temperature and precipitation extremes. Studies conducted in the USA this century have observed an intensification of Lyme disease cases from 2001 to 2014 across the northeastern quadrant of the country, which aligns with predictions of vector-borne transmission and infection patterns due to climate change.

The pressure on land through higher temperatures and extreme weather reduces crop yields, including staple food sources for millions across Africa and Asia; maize, rice, coffee and wheat, to name a few.

During the record-shattering summer of 2024 across southern and eastern Europe, prolonged heatwaves and drought with temperatures surpassing 40°C resulted in widespread crop failures.

In southern Romania, sunflower and corn yields plummeted by up to 90%, while wheat losses across countries like Italy, Poland and France ranged from 25% to 80%. Farmers in several regions struggled with potential bankruptcy as agricultural incomes collapsed under these unprecedented conditions.

When it comes to sea level rise and flooding, fluids find the path of least resistance. As oceans warm and ice melts, so sea levels rise. Sea water contaminates drinking water, as well as delicate coastal fresh water ecosystems. It also makes swathes of habitable coastal land uninhabitable.

As said, over 600 million people live in coastal regions that are less than 10 metres above sea level, that's 10% of the global population and this includes major cities and conurbations such as London and New York. Yet this portion of land is very vulnerable to flooding, not only from longer-term sea-level rise, but surges from storms. Even with 1.5 degrees of warming, forecast sea level rises are in the range 26 to 77cm relative to 2005 levels, which puts at least 136 port megacities at risk.

On a grander scale, the impacts of human-induced climate change lead to weaker economies generally. Flooding is now primarily measured in terms of cost (rather than lives) – for developed nations, and this reaches the billions of dollars/euros/sterling. Time at work is lost and healthcare services are overwhelmed due to extreme weather events, which puts a greater burden on vulnerable economies. Environmental pressures on land and sea reduce the productivity of farming and fisheries and at the same time increase population displacements; so, more people gravitate to smaller regions of land where other communities already live.

The IPCC describes climate change as a 'poverty multiplier'; 1.5 degrees of warming could push 100 million people into extreme poverty. Every environmental issue becomes a global problem.

In 2024, global tree cover loss surged to approximately 30 million hectares – a 5% rise over 2023 – with tropical primary forest loss alone reaching 6.7 million hectares, an area that is roughly the size of Panama. In the Amazon, often called the 'lungs of the Earth', deforestation has surged in recent years.

Every human being on this planet needs the Amazon to continue functioning as it has for millennia.

How Is Climate Change Affecting the UK?

Temperature records globally and locally are increasingly broken every month and every year. The global mean temperature curve over the past 100 years now looks like a steep mountainside rather than a set of undulating hills that peak and trough as they have done through the past 100,000 years.

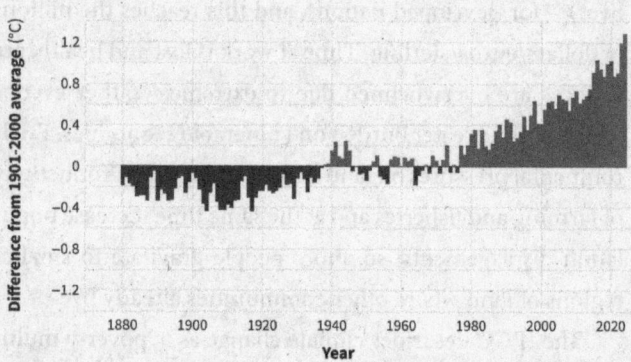

The rise in temperature during the past 100 years

The radical rise in temperature at the end of this graph is unprecedented in climate history. This is what Dr Charles Keeling warned was the 'clear and present danger'. Scientists talk in terms of long-term trends that extend upwards beyond this graph.

CLIMATE CHANGE

Every year, many nations publish their 'The State of the Climate' report. This is an authoritative examination of our changing weather patterns. The Met Office compiles this climate diary for the UK, and also contributes to The State of the Global Climate report.

A major part of the report is focused on extreme weather. By tracking how these patterns are changing, scientists can diagnose the health of our climate system.

In recent years, what is most striking is how exceptional weather is becoming routine.

The 2024 report revealed that the number of days with temperatures 5°C above the 1961–90 average has doubled for the most recent decade 2015–24.

For 8°C above average, it has tripled and for 10°C, it has quadrupled.

The hottest days we experience in the UK have increased in frequency dramatically in just a few decades.

Unsurprisingly, the number of cold nights across a year has dropped dramatically.

Britain's warming climate brings with it a notably wetter character, with recent analysis revealing that enhanced precipitation stems from winter seasonal changes spanning October through March. The winter months of 2015–24 now deliver 16% more rainfall than the 1961–90 reference period across the United Kingdom, highlighting how seasonal precipitation patterns are shifting dramatically in tandem with rising temperatures.

Beyond temperature and rainfall extremes, the report

reveals multiple indicators of Britain's evolving climate. Air and ground frosts have declined steadily since the 1980s, with over two weeks fewer air frosts annually in 2015–24 compared to 1931–90.

Sea surface temperatures now average 0.3°C higher than a decade ago and nearly 1°C above 1961–90 levels. Half of the UK's ten highest coastal sea temperature years occurred in the past decade. This warming is only going in one direction.

Snow events have decreased in frequency and severity since the 1960s, while sunshine hours have increased since the 1980s, primarily driven by brighter winter and spring conditions.

This is consistent with what has happened with global temperature trends generally.

A one-degree rise in global temperature is a substantial increase in energy, manifesting not only in heat but also in extreme wet weather events.

How Do We Reverse Human-Induced Climate Change?

There are as many unknowns as there are knowns when it comes to climate change, but global consensus among most politicians, environmentalists and scientists is that action for deep reductions in greenhouse gas emissions is key.

Most of the emissions of human-induced greenhouse gases come primarily from burning fossil fuels – coal, hydrocarbon gases, natural gas and oil (including petrol and diesel).

For decades, CO_2 emissions have risen in tandem with

fuel consumption, laying the groundwork for economic growth and lifting millions out of poverty. This created a seemingly unbreakable bond between prosperity and carbon emissions – higher GDP per capita invariably meant higher CO_2 emissions per capita.

Yet we now stand at a transformative threshold. The energy sources we choose and the efficiency we achieve in their use hold the power to shatter this historical correlation entirely. Clean energy technologies offer us the unprecedented opportunity to decouple economic prosperity from environmental destruction, creating a future where thriving economies and a stable climate can coexist. The transition from fossil fuels to renewable energy isn't merely an environmental imperative – it's the key to unlocking sustainable abundance for generations to come.

Clean Energy Revolution

While this book focuses on weather, the clean energy transition – a global shift from fossil fuels, such as coal, oil and natural gas, towards low-carbon and renewable energy sources like solar, wind, hydro, geothermal and sustainable bioenergy – provides a crucial context for how climate science influences energy and infrastructure decisions. The 2010s marked the beginning of significant growth in renewable energy sources, as solar and wind power expanded rapidly, while battery storage emerged as a critical component for maintaining power supply during periods of low renewable energy generation.

The clean energy transformation has been decades in the making, but the 2020s have witnessed dramatic scaling. Production costs plummeted, while renewable energy prices reached parity with, and often undercut, those of fossil fuels. The electrification of industry and transport and the adoption of domestic solar, wind and biomass technologies, represents a fundamental shift that many consider irreversible. Along with the continued use of hydroelectric and nuclear power, many countries are now reporting a fall in their greenhouse gas emissions. Although the path ahead is fraught with risk and uncertainty, the progress that has been made brings some degree of hope to future planetary health.

Energy Efficiency Brings Many Rewards
Decarbonising electricity systems is one way of becoming energy efficient. Optimising energy use in buildings through smart meters and architectural designs that naturally ventilate and insulate indoor environments all reduce energy consumption.

Beyond individual efficiency measures, flexible consumer markets are transforming how people interact with energy. Smart technology now enables consumers to shift their electricity usage to off-peak hours when renewable generation is high and prices are low, creating financial savings while supporting grid stability. Time-of-use tariffs and demand response programmes empower households to actively participate in the energy transition, turning consumers into partners in managing supply and demand across the electricity network.

Upcycle, Reduce, Reuse

Ocean expeditions reveal an ecological nightmare: plastic waste is overwhelming our seas at an alarming rate. Marine ecosystems are choking on human waste – a brutal testament to our throwaway culture that kills countless sea creatures daily.

Plastic pollution has become a staggering environmental disaster. While recycling remains costly and often ineffective, policy measures like plastic bag fees and restrictions on single-use bottles offer only modest progress. Critics argue that these efforts barely scratch the surface of a crisis that demands zero net emissions urgently.

The solution lies in fundamental changes in consumption. Government pay-per-waste schemes utilise financial incentives, while consistent recycling standards across businesses and homes encourage a collective responsibility. Environmental charities pressure governments and mobilise communities through mass beach clean-ups, demonstrating public determination to transform wasteful habits.

Individual campaigners demonstrate that personal action can shed light on global problems, inspiring the systemic changes our planet desperately needs to overcome the global plastic issue.

Food and Methane

While CO_2 dominates headlines, methane packs 84 times more heat-trapping power. Since 1750, global methane levels have doubled due to oil and gas extraction, plastic waste and intensive farming.

Livestock generates 14.5% of global greenhouse gases, with cattle being the primary source of this methane surge. This isn't nature's fault – it's our meat addiction driving industrial farming. Solutions exist: plant-based diets benefit both human health and planetary survival, while feeding livestock seaweed, onions and probiotics slashes cow methane emissions by 50%.

Reforestation

Forests: without them, we suffocate.

Industrial agriculture has decimated these carbon-capturing ecosystems, destroying our most powerful climate allies. Beyond CO_2 absorption, forests drive the water cycle, support biodiversity and act as thermal insulators that protect understory life from temperature extremes.

Yet climate change itself is undermining forests' ability to help us. Since 1982, 86% of land ecosystems globally have become progressively less efficient at absorbing CO_2. Groundbreaking research on Amazonia reveals an even more alarming reality: some parts of the Amazon have already become a net emitter of carbon emissions, meaning it releases more carbon into the atmosphere now than it absorbs.

While renewable technologies are crucial, forests and

wetlands remain irreplaceable for cleaning our air and water. Time is running out. The message is clear: the global emergency of climate change defines our generation's greatest challenge.

To reiterate the message from these final paragraphs in the words of the United Nations eighth Secretary-General Ban Ki-moon:

> Saving our planet, lifting people out of poverty, advancing economic growth ... these are one and the same fight. We must connect the dots between climate change, water scarcity, energy shortages, global health, food security and women's empowerment. Solutions to one problem must be solutions for all.

The word petrichor – that distinctive, almost sacred scent that rises from parched earth as the first raindrops fall – captures something fundamental about our relationship with the natural world. It's the smell of renewal, of hope after drought, of nature's promise that even in the darkest moments, life finds a way to begin again.

As we stand at this critical crossroads in human history, facing unprecedented environmental challenges, we must remember that we too are part of this ancient cycle of renewal.

The same human ingenuity that created our climate crisis can engineer our salvation through clean energy, regenerative agriculture and restored ecosystems.

The same hands that cleared forests can plant new ones.

The same minds that built cities can design them to breathe with the seasons.

When our children and grandchildren step outside after summer rain and breathe in that unmistakable fragrance of petrichor, may they inherit a world where humanity has learned to dance with nature's rhythms rather than fight against them – a world where the simple pleasure of rain's sweet scent reminds us not of what we've lost, but of everything we've managed to save.

GLOSSARY

Advection – the horizontal transfer of heat or energy within a fluid.

Advection fog – occurs most typically over a body of water such as a lake or sea. Warm, moist air moves in over the top of relatively cooler air. The warmer air aloft cools to the point where it can't hold the moisture and forms water droplets.

Air mass – a body of air that has similar properties of moisture and temperature. Understanding air masses is a good first indicator of surface conditions and how weather will develop. Air masses are categorised in terms of their source: over land or sea, over a cold or warm region.

Albedo – the ability of a surface to reflect light. Snow has a high albedo, grass has a much lower one.

Altocumulus Lenticularis – type of mid- to high-level cloud that is smooth and shaped like a lens.

Altostratus – medium-level layer cloud.

Asteroid – rocky body that orbits the Earth, asteroids are smaller than planets but still have a significant size.

Atmosphere – the layer of gases above a planet that is held in place by the gravitational force of that planet.

Atmospheric pressure – the weight of air that exerts a pressure on the surface of Earth. Measured in Hectopascals (hPa) or millibars (mb).

Azores High – a semi-permanent area of high pressure that sits over the southern region of the North Atlantic and in particular, close or over the Azores islands. The presence of this anticyclone means low pressure systems track around its periphery.

Benguela Current – ocean current that flows northwards close to the west coast of Southern Africa and is part of the greater circulation pattern of the South Atlantic.

Bergeron process – introduced by scientist Tor Bergeron in 1933, this is a process whereby supercooled water droplets freeze on contact with a solid cloud condensation nuclei, which then grows as water vapour is deposited on them. Main process of the initiation of precipitation.

Blocking High – a persistent zone of high pressure that results in extreme conditions. Very dry for days or weeks, in winter it can mean very cold conditions. It also blocks other weather patterns, which means adjacent regions can receive extreme rainfall.

Blocking patterns – when low and high pressure patterns remain in a similar position for a prolonged period of time, resulting in sustained and similar weather, dry or wet, depending on whether the region is under high or low pressure.

Carbon capture – a geoengineering technique to capture carbon emitted from power stations before the gas is released into the atmosphere.

Carbon cycle – a biogeochemical cycle that describes the exchange of carbon from air, land and sea.

Chlorofluorocarbons (CFCs) – chemicals that were used in refrigerators, aerosols and air conditioning units but have been banned since 1996 as they were found to destroy Stratospheric ozone – a vital component that protects life from harmful UV radiation.

Chromosphere – above the photosphere of the Sun and much hotter, with temperatures rising from 6,000 to 20,000 Celsius.

Cirrostratus – upper-level layer cloud.

Climatologists – scientists who study climate or long-term weather patterns.

GLOSSARY

Cold desert – a dry region of the world where temperatures are very low, such as those found in Antarctica.

Cold front – a boundary between a warm moist air mass ahead and a colder air mass behind. Associated with a change of wind direction, a fall in temperature and a transition to showers/gusty winds.

Condensation – the process by which a gas becomes a fluid as the air temperature falls to a certain level.

Convection – the vertical transfer of heat or energy within a fluid or gas.

Convective Available Potential Energy (CAPE) – a measurement of how much energy is available for convection. The greater the CAPE, the more energised a body of air. A good first indicator as to whether storms from deep convection will produce violent winds, heavy rain, hail and even tornadoes.

Convective zone – a zone within the Sun's interior beyond the radiative zone, where the most dominant process is convection. This process allows heat to travel to the surface of the Sun, and beyond.

Coriolis force – a force that acts on a fluid on a rotating surface, veering the flow to right in the Northern Hemisphere and left in the Southern Hemisphere.

Corona – plasma that extends millions of miles from the surface of the Sun.

Coronal Mass Ejections (CME) – a surge of plasma that is released explosively from the Sun's corona. It can sometimes follow a solar flare. CMEs tend to be associated with sunspots and can extend out into space, attracted to other magnetic fields, including Earths.

Crepuscular rays – optical illusion which appears as shafts of sunlight originating from a point in the sky behind a cloud.

Cryosphere – describes the region of the world covered with ice.
Cumulonimbus cloud – these clouds form due to the process of convection, when the air aloft is much cooler and loaded with moisture. This is largest and most energised type of cloud that produces thunderstorms. The winds can be violent, with heavy rain, hail, thunder and lightning. Tornadoes also develop from the base of the cloud. Cumulonimbus clouds have a distinctive anvil-shaped top to them, which results from the spreading out of cloud as it hits the top of the troposphere.
Cyclone – a large-scale rotating storm system that has a low pressure centre and wind speeds of more than 32mph or 37mph (dependant on the measuring weather centre). Depending on its location and strength, it can categorised as tropical storms, hurricanes and typhoons.
Dew point – the air temperature at which below that point will condense into water droplets. When the dew point and air temperature are the same, there is a high likelihood of fog.
Diurnal temperature range – the range of temperature from night to day accommodating the minima and maxima over a day.
Doldrums – describes weather conditions across the equatorial zone of the Atlantic Ocean; notorious for calm conditions that can unpredictably transition into storms and violent winds.
Eccentricity – the shape of a planet's orbit around the Sun, from circular to elliptical.
Ekman Spiral – describes the action of water when a surface wind blows across it. The Coriolis force results in the surface body of water, or surface current, flowing at 45 degrees to the left of the wind direction in the Southern Hemisphere and consequently the mass transportation of water at depth flows at 90 degrees to the wind direction. This fanning out of the passage of water is observed as a spiral and is one of the processes involved in 'upwelling' across the eastern side of the South Pacific. In the

GLOSSARY

Northern Hemisphere, water flows to the right of the wind direction.

El Niño – a period of warming of the Pacific waters close to the western side of South America, which will lead to a reversal in the trade winds along the equatorial Pacific. It brings a change in weather patterns around the world and is part of what is known as the El Nino Southern Oscillation (see below).

El Niño Southern Oscillation (ENSO) – the shifting of a pressure pattern associated with an El Niño event, when the normally high pressure scenario becomes low pressure across the eastern portion of the South Pacific and results in pressure building to the west of this ocean. This results in drier than average conditions for NE Australia and SE Asia and can lead to heatwave and drought.

Electromagnetic (EM) radiation – radiation emitted by the Sun that consists of a spectrum of wavelengths, from longwave radiation in the form of radio waves to the shortest radiation in the form of gamma rays. Ultraviolet radiation and infrared radiation are all part of the spectrum.

Enhanced Fujita Scale (EFS) – a numerical scale on the strength of a tornado, assessed by observing damage caused by the tornado, which then indicates the strength of the tornado winds.

Ensemble Forecasts – a set of numerical weather prediction forecasts with different initial conditions.

Environmental Modification Convention (ENMOD) – set up in 1978, after the clandestine Operation Popeye, where countries aren't allowed to 'militarize the weather'.

Equator – central line that circumvents the Earth at zero degrees latitude.

Equinox – the time of the year when the Sun is over the equator, resulting in roughly equal daylight hours to night-time. This happens during the spring and autumn.

Evapotranspiration – the combined processes of transpiration and evaporation whereby water from trees, plants (transpiration) and the land (evaporation) is transferred into the atmosphere as it transforms from liquid to a gas (water vapour).

Exosphere – the outer shield of the atmosphere that is exposed to outer space. This outer layer eventually becomes space as its weaker and light molecule and atom content fades to nothing.

Ferrel Cell – one of three atmospheric circulations that describes the three-dimensional grand movement of air across the globe. The Ferrel cell is positioned over the mid-latitudes and sits to the south of the Polar cell and north of the Hadley cell.

Flash flooding – when heavy rain results in surface water accumulating on land, or when a river quickly bursts its banks due to too much water in a short period of time. Flash flooding tends to peak quickly and then ease.

Foehn Wind (Foehn Effect) – A warm wind that flows down the leeward side of a mountain and whose properties have been modified from the windward side. The air is drier, warmer and there are less clouds. As the air descends down the leeward side the added effect of compression warms the air further. So the difference in temperature either side of the mountain can be significant.

Gamma rays – the shortest wavelength in the electromagnetic (EM) spectrum emitted by the Sun.

Geosmin – organic compound that gives off an earthly smell.

Geothermal energy – sourcing heat energy from natural heat underground.

Glacial periods – interval of time during an Ice Age when temperatures are well below seasonal average and describes glacial advances. Interglacial periods occur between glacial periods when Earth is much warmer.

GLOSSARY

Gravitational force – a force that occurs between two bodies. There is a gravitational force between each planet and between the planets and the sun. The bigger the object, the greater their gravitational force. Each body attracts another.

Greenhouse Gases (GHG) – gases in the atmosphere that absorb and re-emit heat. Primary GHG are carbon dioxide, methane, water vapour and nitrous oxides.

High Pressure – a zone of air that is descending in the atmosphere before flowing outward at the surface and circulating around a centre. Indicative of settled weather but strong winds can occur on its periphery and cloud can get trapped within the high cell. In the Northern Hemisphere this pressure system flows clockwise, and conversely, in the Southern Hemisphere.

Hill fog – like advection fog, air is forced to cool and condense due to uplift from the hill.

Humboldt Current – cold ocean current that spreads north from Antarctic across the eastern South Pacific, affecting the western coastline of South America.

Hurricane – a cyclone that develops across the tropical Atlantic, NE Pacific Ocean and Caribbean Sea.

Hydrological cycle or water cycle – the passage of water in all its states of gas, liquid and solid as it travels around the world, from clouds to rivers, oceans and ice.

Infrared – a part of the EM spectrum of radiation emitted by the sun that produces heat.

Inter tropical convergence line (ITCZ) – a zone of unstable air and wet tropical rain, where two air masses converge. It shifts north and south annually depending on the position of the sun and enhances energy and momentum to seasonal rains across the globe, including the Indian monsoon.

Ionosphere – an outer layer of the Earth's atmosphere 85 to 1,000km above the surface. This is where extreme ultraviolet radiation and X-rays from the Sun ionise the atoms and molecules.

Jet stream – a fast-moving stream of air in the upper troposphere that initiates development of weather at the surface depending on its strength, shape and position. The jet stream across the Atlantic Ocean develops and drives weather towards Europe. A similar set-up occurs across the midlatitudes around the world.

Keeling Curve – Depicts the rise in CO_2 at Mauna Loa, an observation site in Hawaii whereby CO_2 has been increasing annually since observations began in 1958 by scientist Dr Charles Keeling. It also shows a seasonal variation in levels.

La Niña – significant cooling of the Pacific waters along the Western side of South America. This part of the southeast Pacific is normally cold but further 'upwelling' from deep seawater reduces the temperature further. It can result in a change of normal weather patterns around the world.

Lee gustiness – gusts that form during stable air flow. When a wind hits a mountain and gets deflected, rotors of air are then created; on the lee side of the mountain and wind gusts can strengthen.

Leeward – the sheltered side of a mountain range or set of hills. Weather conditions tend to be warmer and drier, but winds can turn quite gusty.

Lifting Condensation Level (LCL) – the height in the atmosphere, where the humidity of a rising pocket of air reaches 100%. This is where the cloud base will form.

Low pressure – a circulation of air that spirals inwards and upwards, associated with strong winds and outbreaks of rain. In the Northern Hemisphere the air travels in an anti-clockwise direction, the converse is true in the Southern Hemisphere.

Mesosphere – the third layer of the atmosphere directly above the stratosphere. It extends from about 50 to 80km above sea level, where temperature decreases with height. Meso means middle and this is the middle layer of the Earth's atmosphere.

GLOSSARY

An important layer for burning up debris from space, including meteors.

Meteor – a meteoroid that burns up as it passes through the Earth's atmosphere, seen as a streak of light or a shooting star.

Meteorite – a meteoroid that survives the passage through the Earths atmosphere and makes impact on Earth.

Meteoroid – rocky, the body sometimes contains metal that travels through space. Sizes range from specks of dust to ten metres in diameter so smaller than an asteroid.

Meteorological Office, Met Office – the UK's national weather service

METOCs – weather forecasters from the Royal Navy who specialise in meteorology and oceanography.

Mid-latitudes – the temperate zone of the world between 30 and 60 degrees north or south, where the climate tends to transition through four seasons. Unlike the tropics, where there is a wet and dry season.

Mistral wind – a local wind to France and the Mediterranean. An intense violent and cold north or north-westerly wind that often follows a cold front. The air is very clear, with excellent visibility.

Moore's Law – theoretical growth in computing power suggests a doubling every 18 months.

Neap tides – minimal tidal range; when the sun and moon oppose each other, resulting in a weaker gravitational pull on Earth. Occurs twice in a lunar month.

Near Earth Objects (NEO) – any object in space that is in close proximity to Earth.

Nimbostratus – rain-bearing layer cloud in the lower atmosphere.

Nimbus cloud – nimbus derived from the Latin meaning 'rain bearing'. These are thick grey clouds, with a straggled appearance, that produce drizzle or light rain.

Noctilucent clouds – rare clouds that form in the mesosphere and are only observed during twilight across the higher latitudes of Earth.

Nuclear energy – an energy source from nuclear reactions of plutonium and uranium contained within a nuclear power plant. Described as a low-carbon energy.

Nuclear winter – a lengthy spell of colder weather following a nuclear holocaust, where the huge volumes of dust and ash from the fires of detonated atomic bombs remain in the atmosphere and block out the sun for years. The climate impacts are suggested to be far-reaching, mainly colder and much drier.

Numerical Weather Prediction (NWP) – uses mathematical equations and atmospheric/oceanic models to predict future weather scenarios based on current conditions.

Obliquity – the angle of tilt of a planet. The greater the tilt, the more extreme the planet's seasons.

Ocean buoys – floating vessels that measure ocean and weather data.

Omega block – a type of atmospheric blocking pattern whereby the high pressure and adjacent low pressures either side of the high create a shape that looks like the Greek capital letter Omega.

Operation Cumulus – UK military experiment to seed clouds. Some suggest it resulted in a flash flooding in Lynmouth, Devon, in 1952, when 35 people died. It was never conclusive.

Operation Popeye – a geoengineering project during the Vietnam War to enhance rain through cloud seeding over the Ho Chi Minh Trail.

Ozone layer – a very thin part of the stratosphere that absorbs much of the ultraviolet radiation from the Sun, 90% of the UVB and all of the UVC, the shortest of three UV wavelengths. UVA is able to penetrate through the ozone layer.

GLOSSARY

Paris Agreement – a climate agreement within the United Nations Framework Convention on Climate Change (UNFCCC), signed by 196 member countries in December 2016 to keep the global temperature below two degrees of warming, in an effort to reduce the impact of climate change.

Peasouper – a thick and noxious fog found during the industrial era in large cities when the atmospheric conditions were very calm and there was no wind or rain to clear the air.

Percolation – the process by which water is absorbed by soils, or seeps into the rocks, that sometimes then arrives in underwater lakes and aquifers.

Permafrost – ground or soils that is permanently frozen.

Petrichor – the name given to the smell of the air when it has rained, after a dry spell.

Photosphere – the visible surface of the Sun, where light is radiated out.

Plasma – the hottest state of matter, the coldest being a solid, the second liquid. The third is gas, the fourth is plasma. The Sun is hot enough to produce plasma and this consists of highly charged particles. These particles are so energised that they defy the Sun's gravitational field and escape into space.

Precession – the wobble that occurs during the spin of a planet.

Precipitation – anything that has condensed in the atmosphere and then falls through the sky due to gravity, such as rain, snow or hail.

Radiation fog – a common type of fog that occurs under clear skies at night and with little wind, the air will cool and condense into tiny water droplets.

Radiative zone – a zone within the Sun's interior, where the most dominant process is radiation. It exists around the core of the Sun. Energy in this zone is produced by nuclear fusion.

Radio waves – a long-wave form of radiation that is part of the electromagnetic radiation emitted by the Sun.

Rain shadow – a dry zone that sits to the lees of hills or mountains.

Ridges – peaks in the upper pattern of winds that is associated with surface high pressure and settled weather.

Rossby Waves – also known as planetary waves, Rossby Waves are an upper-wave pattern of air that hosts a number of jet streams. They can be seen propagating the Northern and Southern Hemispheres across the midlatitudes and are driven by the temperature gradient from the poles and tropics, the Coriolis force and the conservation of vorticity.

Saharan Air Layer (SAL) – hot dust layer of air that extends from West Africa across the Atlantic and sits above a cooler moisture layer of air that has a strong ocean influence. The presence of this layer significantly reduces the development of weather systems with height.

Savannah – a biome of mostly grassland, shrubs and very few trees. The lack of rain doesn't support forests. It is found between deserts and rainforests.

Solar cycles – the Sun's magnetic activity goes through periodic peaks and troughs. The overall cycle from weaker activity to its strongest flares is about 11 years, but this varies over centuries.

Solar flares – large eruptions of radiation and plasma that explode from the sun. Seen as bright flashes of light.

Solar radiation – energy emitted by the sun in the form of a spectrum of electromagnetic radiation.

Solar storm (geomagnetic storm) – a storm wind that surges towards Earth with huge explosive momentum and dives towards Earth's magnetic poles.

Solar wind – a stream of highly charged particles, mainly protons and electrons, in a plasma state that extend and are released from the Sun's corona.

GLOSSARY

South Atlantic Gyre – large ocean circulation pattern across the South Atlantic.

Spring tides – the greatest tidal range, when the sun and moon are aligned, which results in a strong gravitational pull on Earth. Occurs twice in a lunar month.

Storm surge – a surge of water, significantly higher than normal, which is the result of stormy conditions whereby air pressure is lower, allowing sea level to rise, and where strong winds push the body of water in a particular direction. This can cause coastal flooding when the surge travels in the direction of land.

Stratocumulus – a low-level cloud that has some definition of convection but remains structured in a layer form. A common cloud that doesn't tend to produce rain.

Stratosphere – the second layer of Earth's atmosphere that warms with height. This is where the ozone layer can be found and the reason why this part of the atmosphere is hotter.

Stratospheric Polar Vortex (SPV) – the strong vortex of wind over the poles at 50km, which develops over the polar winter months and captures the colder air within its circulation.

Stratus – a layer cloud in the lowest part of the sky. Stratus is derived from the Latin 'layer'.

Sudden Stratospheric Warming (SSW) – the sudden weakening of the stratospheric polar vortex over the poles during winter that leads to a slackening of previously strong persistent winds high in the atmosphere and ultimately leads to a reversal of winds. This can have consequences at ground level weeks or months later, with weakening of jet streams and a dominance of other winds. Over the UK and Europe, it can lead to much colder ends to the winter.

Super Typhoon Haiyan – a deadly typhoon that devastated the Philippines in 2013. With sustained winds of 195mph, it killed over 6,000 people.

Supercooled water – a liquid state of water where the temperature is below freezing, it can be one of the main ingredients for cloud and raindrop formation.

Thermosphere – the fourth atmospheric layer of Earth and higher than the mesosphere. The temperature increases with height due to the absorption of radiation from the Sun, and varies depending on solar radiation.

Tornado – an intense vortex of wind that develops from cumulonimbus clouds. The strongest winds that affect the surface of the Earth, they can cause much damage.

TORnado and storm Research Organisation (TORRO) – UK based research group focusing on storms and tornadoes.

Trade winds – a set of persistent and strong winds between 30 degrees north and south, found in the tropics blowing towards the equator. Blowing from the northeast in the northern hemisphere, and blowing from the southeast in the southern hemisphere.

Tropical Depression – a large area of low pressure located over the tropics and sometimes a precursor to cyclone development.

Troposphere – the lowest layer of the atmosphere, where temperature cools, with height resulting in cloud formation and subsequent weather. This is the zone of the atmosphere where the hydrological cycle allows water to replenish in all its forms.

Troughs – dips in the upper pattern of winds associated with surface low pressure and unsettled weather.

Ultraviolet radiation – part of the electromagnetic spectrum emitted by the Sun. Most is absorbed by the ozone layer in the stratosphere, but some UV radiation penetrates to the Earth's surface, causing sunburn and skin cancer.

Walker cell – a theoretical atmospheric circulation used to describe the rising and falling of the air across the tropics that extends south across the South Pacific. It is used as part of the explanation of the El Niño–Southern Oscillation (ENSO).

Warm front – indicates a change of air mass to a warmer, moisture-laden body of air that can produce thicker clouds, higher humidity and rain or drizzle.

Water vapour – a greenhouse gas made up of water molecules (H_2O), the gaseous state of water which occurs when liquid water reaches boiling point.

Waterspout – a vortex or a swirling wind that extends from a body of water up to the base of a cloud. Looks like a tornado and in some cases is formed the same way as a tornado.

Wavelength – the length between either two crests or troughs in the wave pattern emitted by the spectrum of colours. In the visible light spectrum, violet has short wavelength up to red, with a longer wavelength.

Weather or sounding balloons – a balloon that carries weather instruments, called radiosondes. It rises through the atmosphere to measure the humidity, pressure and temperature. Wind speed and direction is derived from GPS.

Weather satellites – measures weather and climate data from space; some are stationary, others orbit the Earth.

Wind shear – the change in strength and direction of wind with height. This enhances some atmospheric processes and reduces the potency of others.

Windward – part of a mountain range or set of hills exposed to a prevailing wind; weather conditions tend to be more raw, with stronger winds, more cloud and chance of rain.

Winter Solstice – occurs when the Earth's poles are furthest from the sun, or at the furthest point in the other hemisphere.

X-rays – a type of radiation emitted by the Sun, which has a very short wavelength.

ACKNOWLEDGEMENTS

This book wouldn't have been possible without the fascination from our editor Madiya Altaf. Thank you for all your support and guidance, and for listening to us when we said there's so much more to the weather! A huge thank you to Vickie White, the most amazing and dedicated agent. Vickie, without your powerhouse of inspiration and energy we would not have united with Bonnier Books UK and written this book.

Simon: I am eternally grateful for the support my wife Emma has given, especially during the writing of this book. All the time she's spent entertaining Noah and Nell to give me space while I've been writing is really appreciated, thank you.

My fascination with the weather started at an early age so I'd like to thank my parents and teachers at school for embracing my weather obsession, the staff at the Meteorology Department at the University of Reading for your brilliant tuition and to the Met Office and BBC for giving me my dream career.

Clare: Thank you my amazing family who always let me wax lyrical about weather as a child. To the scientists and forecasters at the Met Office – their passion, experience and scientific diligence is a true inspiration to me. Thank you also to my meteorological mentors, Prof. Phil Dyke, Jim Bacon, Margaret Emerson and Rob Varley.